Soil Fertility and Nutrient Management

About the Author

 Dr. Sharanappa obtained his B.Sc.(Agriculture), M.Sc. (Agriculture) and Ph.D. (Agronomy) from University of Agricultural Sciences Bangalore. He served in the University of Agricultural Sciences as Assistant Professor, Associate Professor, Professor and Professor and Head of the Department of Agronomy for 34 years in teaching, research and extension. At present as Emeritus Professor of ICAR. Under his guidance nine students secured PhD. (Agronomy) and 14 students M. Sc.(Agriculture) in the University of Agricultural Sciences, Bangalore. He served as member of the advisory committee of 68 postgraduate students. His special interest in teaching and research are Organic Farming, Soil Fertility and Nutrient Management in Crops and Cropping Systems. He has developed technologies for nutrient management for organic groundnut, onion, chilli, baby corn and finger millet production. He served as the Principal Investigator for DST funded research project on establishment of organic farming technology park in agroclimatalogically disadvantaged region of Karnataka state and NPOF funded project on establishment of Model Organic Farm at GKVK, Bengaluru. Thirty three research papers were published by him in NAAS rated scientific journals, and published the book on Oilseed Production. He has presented research papers in the international conferences held in Israel and USA on the precision nutrient management in maize and groundnut.

Soil Fertility and Nutrient Management
Principles and Practices

Sharanappa
Emeritus Professor
Department of Agronomy
College of Agriculture
University of Agricultural Sciences
GKVK, Bengaluru, Karnataka

CRC Press
Taylor & Francis Group
Boca Raton London New York

CRC Press is an imprint of the
Taylor & Francis Group, an **informa** business

NEW INDIA PUBLISHING AGENCY
New Delhi-110 034

First published 2023
by CRC Press
4 Park Square, Milton Park, Abingdon, Oxon, OX14 4RN

and by CRC Press
6000 Broken Sound Parkway NW, Suite 300, Boca Raton, FL 33487-2742

© 2023 New India Publishing Agency

CRC Press is an imprint of Informa UK Limited

The right of Sharanappa to be identified as author of this work has been asserted in accordance with sections 77 and 78 of the Copyright, Designs and Patents Act 1988.

Print and electronic editions not for sale in South Asia (India, Sri Lanka, Nepal, Bangladesh, Pakistan, Afghanistan and Bhutan).

British Library Cataloguing-in-Publication Data
A catalogue record for this book is available from the British Library

ISBN: 9781032429298 (hbk)
ISBN: 9781032429304 (pbk)
ISBN: 9781003364955 (ebk)

DOI: 10.4324/9781003364955

Typeset in Times New Roman
by NIPA, Delhi

Preface

Sustainable agriculture productivity depends on successful maintenance of soil fertility. Among the 16 essential elements required by the plants carbon, hydrogen and oxygen are taken from air and water which accounts for about 96 per cent of the plant composition while the rest account for about 4 per cent called mineral nutrients. These are absorbed by the plants from soil. They play structural and functional role in the plants, besides there are some elements which play beneficial role in the plants. The mineral elements interact with soil organic matter, clay minerals, soil microorganisms and other associated mineral elements. These interactions determine their availability and dynamics in the soil. Understanding of the dynamics of plant nutrients in the soil will provide scientific basis for efficient nutrient management. Soil organic matter not only provides the nutrients required by the crop but also improve the biological and physical properties of the soil. Attempt has also made to provide information on production and management of organic manures, biofertilizers, integrated nutrient management in cropping systems and nutrient management in problematic soils. The author hope that this text will serve as a valuable guide for students and teachers for learning and teaching respectively on soil fertility and nutrient management in crop production.

Sharanappa

Contents

Chapter 1

Soil Fertility and Productivity Concepts and Factors

ESSENTIAL ELEMENTS FOR PLANT GROWTH

Arnon (1954) set three criteria to state an element is essential to plants. They are:

1. The plant must be unable to grow normally or complete its life cycle in the absence of the element,
2. The element is specific and cannot be replaced by another and,
3. The element plays a direct role in metabolism.

Accordingly there are 16 elements essential for plant growth. They are carbon, hydrogen and oxygen called primary nutrients, calcium, magnesium and sulphur called secondary nutrients; iron, zinc, manganese, copper, boron, chlorine and molybdenum are known as micronutrients.

Table 1. The average composition of essential elements in plants.

Element	Composition (%)	Element	Composition (ppm)
H	6.0	Cl	100
O	45.0	Fe	100
C	45.0	B	20
N	1.5	Mn	50
K	1.0	Zn	20
Ca	0.5	Cu	6
Mg	0.2	Mo	0.1
P	0.2		
S	0.1		

Nicholas (1963) suggested 'functional nutrients' in preference to micronutrients. For all such elements which play a role in metabolism but requirement of which can be met by another element of function or which can be by-passed by supplying the product of the reaction mediated by the elements.

Soil fertility: Soil fertility is the ability of the soil to supply all the essential nutrients in optimum amount in balanced proportion in a form readily available to plants under conditions favorable for plant growth and should be free from toxic substances.

Soil fertility has been considered in the past in restricted sense as a physicochemical phenomenon or as an index of available nutrients for the plants but the modern usage of the soil to produce crop of economic value to man and maintain the health of the soil to future use.

Even though the soil fertility is the inherent property of a given soil, the plants can modify soil fertility in two ways.

1. Rhizosphere effect is exerted by the plants which can alter the fluxes of energy and the supply of substrates for microorganisms.

2. Inherently different growth rates and metabolism of different plant species that are known to change the capacity of the soil to provide each particular plant with nutrients or other words ability of the plant to exploit the nutritional supply of a given soil depends on type of rooting characteristics (deep/shallow), rates and patterns of exudations.

Mineral fertility of the soil: It refers to the good supply of cations and anions requirement by the plant. It is only one part of the general natural fertility of the soil in which physical, chemical and biological properties are considered.

The human effort through ploughing, addition of fertilizers and various agrochemicals adds to the natural fertility known as acquired fertility.

Soil fertility qualities used without the qualities of 'biological', 'physical' or chemical gives insufficient information about the state of soil. These prefixes allow interpretation to be focused on components or combinations of components of soil fertility that are influenced by management decisions.

Soil biological fertility or soil biological health is used within the frame work. Unfortunately, there are no simple widely applicable and quantitative measures of any aspects of soil biological fertility because they are constrained by parent rock, soil origin, land scape. Of late when the problems of conservation of agricultural soils became crucial a new view of fertility emerged. This approach integrates the nutritional needs of the human species and those of living organisms. It broadens the notion of 'producer of crops' to a duration that guarantees the edaphic conditions of the entire biocoenosis. It incorporates the sustainability aspect and importance of living organisms and their actions on soil fertility.

In Switzerland soil fertility has been defined officially in an ordinance on soil pollutants (1998) encompassing all the aspects as follows.

The soil is considered fertile when

• It has a diversified and biologically active biocoenosis, structure typical for the site and intact degradation capacity.

- It enables plants and plant associations natural or cultivated to grow and develop normally without being injurious to its properties.
- The fodder and plant products it provides are of good quality and do not threaten the health of man and animals.
- Its ingestion or inhalation do not endanger the health of man and animals.

Soil productivity: Soil productivity is defined as the capability of soil for producing a specified crop or sequence of crops under defined set of management practices. It is measured in terms of outputs or harvests in relation to the inputs of production factors for specific kinds of soils under a physically defined system of management.

Soil Health: The continued capacity of soil to function as vital living system, within ecosystem and land use boundaries, to sustain biological productivity, maintain the quality of air and water environments and promote plant, animal and human health.

The physical, chemical and biological indicators for screening the condition, quality and health of the soil are:

1. Texture
2. Depth of soil
3. Infiltration rate
4. Soil bulk density
5. Water holding capacity
6. Soil organic matter
7. Total organic carbon
8. Nitrogen
9. Soil reaction (pH)
10. Electrical conductivity (EC)
11. Extractable NPK
12. Microbial carbon and nitrogen
13. Potential mineralizable nitrogen
14. Soil respiration
15. Temperature

Organic matter indices of soil productivity are:

- Total soil organic matter (TSOM)
- Microbial biomass soil organic matter (MBSOM)
- Ratio of MBSOM/TSOM

- Labile soil organic matter (LSOM)
- Ratio of LSOM/TSOM
- N mineralization capacity
- Charge contribution to soil
- Quantity and quality of organic inputs used
- Soil fauna and microflora – trends in earthworms, nematodes, termites, mites and collembola from cropping cycle to cycle.

Soil quality and Soil health are synonymously used. Soil quality is defined as capacity of the soil to function within ecosystem boundaries and interact positively with the environment external to that ecosystem.

A quality soil performs three functions satisfactorily. These are also indicators of soil quality

- Provide medium for plant growth
- Regulate water flow in hydrological units
- Serve as environmental filter

Soil Quality Rating (SQR) has three sub factors

$$SQR = OM + TP + ER$$

- **Organic matter (OM):** Includes all decomposed organic material from plant sources either grown or imported to the site and returned to the soil.
- **Tillage practices (TP):** All field operations which break down residues and aerate the soil including tillage, planting, fertilizer injections *etc.*
- **Erosion(ER):** Ssoil erosion is the removal and sorting of surface soil material by erosion.

Soil quality rating is used as a tool to evaluate farming practices which maintain or increase soil organic matter and enhance the biological process that improved physical condition of the soil.

Factors influencing the soil fertility

Physical factors: The factors of soil formation influence soil fertility

$$\text{Soil formation} = f(P\ C\ T_o\ T\ L_o)$$

Parent material: soil formed from sand alone or quartz is poor in nutrients. Apatite and feldspar are rich in phosphorus and potassium respectively. Calcite is rich in calcium, dolomite rich in calcium and magnesium, soils from gneiss and schist are fertile, volcanic soils like Andesols are rich in nutrients.

Climate: Soils of low rainfall areas are rich in bases viz. calcium, magnesium and sodium. Soils in high rainfall areas are rich in iron, manganese and aluminium. In these areas organic matter production and addition is greater. In arid and semi-arid areas soils are poor in organic matter due to faster decomposition and less addition. Soils in extremely dry deserts are youthful with poorly defined horizons. As one proceeds from the drier region to high rainfall region soils get deeper and more leached with a distinctive 'A' horizon and definite 'B' horizon. As rainfall increases the 'C' horizon is found at deeper depths.

Temperature: High temperature with low rainfall leads to less leaching and accumulation of cations and anions in the soil viz. Ca^{++}, Mg^{++}, HCO_3^-, CO_3^- and SO_4^-.

Topography: Modifies the effects of parent material and climate. Accumulation of eroded soils in valleys enriches the native soil with nutrients. Soil in upper region in sloppy areas is depleted of soil and nutrients.

Soils of volcanic eruptions are rich in nutrients e.g., Andesols. These have attained equilibrium, less leaching and erosion.

Depth of soils: In deeper soil plants can exploit greater volume of soil, while in shallow soils the volume of soil available for nutrient exploitation is very less.

Presence of hard pans/clods: reduce the volume of the soil available for nutrient exploitation. Nutrients are physically inaccessible to the plants.

Excessive irrigation or water logging leads to leaching of nutrients. Nitrogen is lost by nitrate leaching and denitrification. While in dry soils mobility of the nutrients is poor.

Clay fractions: The order of soil particle size is Gravelly > sandy > silty > clay (< 2 μm). Clay particles due to presence of surface charge are highly reactive and form sheet of ion exchange in soil and thus control and regulate adsorption, retention and release of many plant nutrients.

Clay minerals like kaolinite or halloycite, nacrite and dickite (1:1 minerals) are non-expanding while smechtite, vermiculite or micas (2:1 minerals) are expanding type clay minerals. They have silica (tetrahedral) and alumina or magnesia (octahedral) sheets.

Chlorites have 2:1:1 layers which comprises 2:1 silicate layer alternated with magnesium or aluminium dominated dioctahedra sheet depending on acidity or alkalinity of soils. While there are mixed types (1:1, 2:1 and 2:2) example: smectite- kaolinite, mica- vermiculite, chlorite- vermiculite. 2:1 type of clay minerals have more CEC and can fix ammonium and potassium in their inter lattice layer.

Table 2. Cation exchange capacity of clay minerals and organic colloids

Type of soil	CEC (c mol (P^+) kg^{-1})
Kaolinite	3-10
Mica	10-40
Chlorite	10-40
Montmorillonite	80-150
Vermiculite	100-150
Organic colloids	>200

Chemical factors

Those factors influencing the availability of nutrients affect the soil fertility like fixation of nutrients by formation of insoluble compounds like tricalcium phosphate, ferric phosphates, aluminum phosphate and ferric form of iron. Excess of certain nutrients or imbalance nutrients or presence of toxic elements in soil impairs soil fertility. Mobility of the ions in the soil- is in the order of $NO_3^- > K^+ > H_2PO_4^-$ or HPO_4^-. Presence of soluble salts (*i.e.*, Ca, Mg, Na and K salts of Cl^-, SO_4^-, CO_3^-, HCO_3^- NO_3^- and borates) in soil is measured by electrical conductivity. If the electrical conductivity is > 4 d Sm^{-1} it is harmful for crop growth. Free oxides of Fe, Al, and Si which are coated on the soil particles impairs soil fertility.

In calcareous soils extractable calcium and magnesium exceeds the cation exchange capacity.

In saline soils calcium and magnesium or total soluble salt load is high enough to interfere with the growth of most plants.

Non alkali saline soils: Excessive soluble salts are calcium and magnesium.

Saline- alkali soils: Excessive soluble salt can be primarily sodium.

Saline and alkali soils affect the plant growth in several ways.

High osmotic pressure of soil solution results in decrease in physiological availability of water to plants.

High concentration of some cations depresses the uptake of others which may be essential for plant growth. If excess of calcium and magnesium in soil depress the plant uptake of potassium.

Deficiency iron, zinc, manganese and copper are more pronounced in soils with pH ranging from neutral to basic (7.0 to 8.5) or calcium saturation.

Biological factors: A gram of fertile soil contains billions of microorganisms which approximately weigh about 4000 kg ha^{-1} and may constitute 0.01 to 0.4% of the total soil mass.

Roots of higher plants – macroflora; bacteria, actinomycetes and algae- microflora; Protozoa and nematodes – microfauna.

Type of cropping influences the soil fertility either through the quantity of nutrients absorbed or added through their litter, nodules *etc.*

Microbial biomass bacteria, fungi and microfauna are associated with particulate and colloidal fractions of soil are involved in organic matter dynamics and nutrient acquisition. Earthworms and other macrofaunal community are associated with mixing and pore formation and primary decomposition of the litter.

- Mycorrhiza – external mycelium, spores
- Nitrogen fixation – free living or symbiotic
- Nutrient mobilizers – phosphorus solubilizing organisms
- Decomposers/mineralizers/immobilizers
- The labile soil organic matter

All these are associated with mineral and organic particles, which are approximated by the particulate organic matter which is indicative of short term fertilizer property of soil organic matter.

Even though the soils may contain adequate nutrients required for plant growth under circumstances of inadequate moisture supply, ill drainage, presence of excess amounts of salts, toxic levels of nutrients, extreme textures, shallow rooting depth, and infestation with pest and disease causing organism or problematic weeds (root grub, cut worm, root wilt disease, weeds like *Cynodon dactylon* and *Striga* sp.) are not productive.

QUESTIONS

Define
1. Essential elements
2. Functional nutrients
3. Soil fertility
4. Mineral fertility of soil
5. Biological fertility of soil
6. Soil productivity
7. Soil health
8. Soil quality

Answer the following
1. What are the indicators of soil health?
2. What are the physical factors influencing soil fertility?
3. List chemical factors influencing soil fertility.

4. What is the significant influence of soil flora and fauna on soil fertility?
5. "All fertile soils are not productive." Justify.
6. What are the organic based indicators of soil fertility?

CHOOSE THE CORRECT ANSWER

1. The per cent composition of primary nutrients in plants is
   (a) 70   (b) 96   (c) 4   (d) 45

2. The essentiality criterion of plant nutrients was stated by
   (a) Liebig   (b) Beizerinch
   (c) Arnon   (d) Nicholas

3. One of the following cannot be assessed in the laboratory
   (a) Soil health   (b) mineral fertility
   (c) Soil fertility   (d) soil productivity

4. The inherent fertility of the soil is altered by
   (a) Topography and rainfall   (b) Wind and topography
   (c) Mineral content of the soil (d) Rainfall and temperature

5. Soil rich in one of the following clay mineral hold less nutrients
   (a) Kavolinite   (b) Vermiculite
   (c) Micas   (d) Montmorillonite

6. Acid soils are not fertile due to
   (a) Excess of Fe, Mn and Al
   (b) Poor organic matter content of the soil
   (c) Greater infiltration of water
   (d) Greater activity of fungi

7. Saline soils are not fertile due to
   (a) Excess of salts of Ca and Mg
   (b) Excess of Na
   (c) High osmatic pressure
   (d) Poor drainage

Chapter 2

Nutrient Availability in Soil and Crop Response

Concept of nutrient availability: The ions in the soil solution are available to plant provided they can reach an active root or it can reach them. The concentration of ions in the soil solution at any given moment is however; usually low, except immediately after the application of water soluble fertilizers.

Most of the nutrients are taken up by the plant in the form of cations which are positively charged. Available cations either in the soil solution or held against leaching by negatively charged clay or humus particles are easily taken up by the plants. Other plant nutrients notably phosphorus, boron and molybdenum are taken up by the plant as anions which are negatively charged. The concentration of anions in the soil solution is usually low because they become converted to new less soluble forms *e.g.*, soluble phosphate ions forms less soluble iron or aluminium phosphates. Minerals upon weathering and organic matter upon mineralization release nutrient ions.

Physical factors

However, much readily available nutrients a soil may contain various physical features of the soil make it impossible for the plant roots to absorb it. It then becomes positionally unavailable. These features include hard pans, undesirable soil structures, presence of hard clods, dry soil horizons and water logged situation.

Low soil cation exchange capacity leads to less retention of calcium, magnesium, potassium, copper and zinc.

Copper is strongly bound to organic matter in peat and muck soils, potassium and ammonium are trapped in the inter lattice layer of 2:1 type clay minerals.

Climatic factors

Low soil temperature retards the release of nutrients from organic matter by mineralization. Similarly the release of nutrients by weathering of rocks is also reduced.

Competing factors for nutrients

Because of many competing factors the proportion of the total plant nutrients in the soil taken up annually by the crop is very small. In some cases much less than 1%. Even slowly available water soluble nutrients from fertilizers are subject to leaching. Uptake by weeds and utilization by soil organisms and immobilization of nitrogen by microorganisms when organic matter with wide C: N are added to the soil reduce availability of nutrients to crops.

ROOTING SYSTEM

Shallow rooted crops: nutrient availability is less. Deep rooted crops/crops with ramified root system can utilize nutrients from the greater volume of soil.

Factors restricting availability of nutrients

Salt affected soils: in calcareous soils the phosphorus is fixed as tricalcium phosphate, similarly ammonical form of nitrogen transforms into gaseous ammonia which will be lost by volatilization. Micronutrients *viz.* iron, manganese, zinc, boron copper and cobalt are converted into less soluble forms. Potassium availability is reduced due excess activity of calcium in saline soils and sodium in alkali soils.

Acidic soils: Phosphorus fixed as iron phosphate or aluminum phosphate, reduced mineralization and nitrification, nitrogen fixation by symbiotic and asymbiotic microorganisms. Calcium, magnesium, potassium, copper and zinc are leached from the soil due to high rainfall. Molybdenum is converted into unavailable form. Manganese and aluminium become more soluble and may become toxic to the plants there by restrict the uptake of other nutrients.

Soil reaction *vs* availability of nutrients

At pH < 5 iron, aluminium and manganese become more soluble and toxic. As the pH increases towards neutrality (moderate acidity) their solubility or availability is not so deprived. They will not develop toxic limit. If pH is more (> 7) plants develop deficiency symptoms. Copper, zinc and cobalt are also available at low pH values. At pH 7 they become critical. The availability of nitrogen, sulphur and molybdenum are somewhat restricted at low pH values, phosphorus is more available at intermediate pH values. Calcium and magnesium availability increases with increase in pH except in soils which have appreciable amount of sodium. Even in degraded alkali soils increase in pH shows presence of calcium and magnesium. Boron availability is more in the pH range 5 to 7. At pH < 5, boron availability decreases between pH 7 and 8.5 (between 7.0 and 8.5 there is fixation of boron by calcium which hinder the movement and metabolism of boron), with increase in pH from 8.5, the boron availability increases, molybdenum is not available in strongly acid soils (at pH 4). As the pH of the soil increases (6 and above) availability of nutrients increases. Chlorine is unavailable in alkaline soils. The favorable pH for most of the crops is between 6 and 7.

Antagonism of nutrients: If one or more essential elements are available in excess this will adversely affect the uptake and the efficiency of the other nutrients.

Phosphorus × Zinc: If phosphorus is excess leads to zinc deficiency.

Potassium × Magnesium: Excess of potassium in soil depress the uptake of magnesium and calcium. Potassium is readily taken up by the plants than other elements and is mobile in the plants while magnesium is less mobile becomes deficient. In calcareous soils calcium is high which depresses the uptake of potassium known as calcium induced potassium deficiency in calcareous soils. Similarly sodium induced potassium deficiency in sodic soils. Magnesium induced calcium or potassium deficiency in magnesium rich soils is also common.

Nutrient deficiencies are induced through antagonism are caused by excessive use of lime or neglect of liming where it is absolutely essential, imbalanced use of fertilizer.

In acidic soils excess of H^+, Al^{3+} depresses the uptake of calcium, magnesium and potassium.

Synergism: When two nutrient elements each reinforce the influence of the other on plant growth they are said to be synergetic. The following are some of the examples:

Nitrogen × Phosphorus,

Nitrogen × Sulphur,

Nitrogen × Molybdenum and

Nitrogen × Zinc, Iron, potassium and phosphorus.

Leguminous plants have synergy with *Rhizobia* sp., calcium, phosphorus and cobalt.

Crop response to nutrients

Crop growth is controlled by several growth factors. Deficiency or absence of nutrients in the soil hinders the crop growth and productivity

Liebig (1855) stated that "with equal supplies of atmospheric condition of the growth of plants, the yields are directly proportional to the mineral nutrients supplied in manure." Further stated that crop yield cannot be increased by adding more of the same substance. For a single deficient nutrient with increasing addition of a fertilizer the crop yield may be represented by a straight line. Sloping upward from the control yield to a higher yield at which another factor becomes limiting. At this point, the linear increase ceases, and with further additions of the nutrient, the yield remains substantially constant.

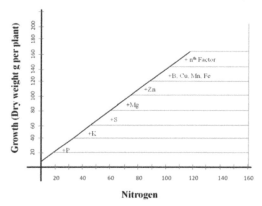

Fig 1. Response of crop to addition of single deficient nutrient in soil.

Mitscherlich (1909) grew oat plants in sand cultures treated with different quantities of phosphorus. He observed that oat yield was not a simple straight line function of the quantity of fertilizer phosphorus supplied. The rate of increase in yield is relatively small at the point corresponding to the greatest application of fertilizer phosphorus, so that the curve seemingly is approaching a maximum beyond which additional quantities of phosphorus would produce no further increase in yield.

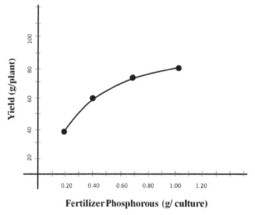

Fig 2. Response of oat crop to addition of phosphorus fertilizer.

This state of affairs suggested to Mitscherlich the concept that the rate of increase in yield produced by adding fertilizer phosphorus is proportional to the decrement from the maximum yield which is mathematically expressed as.

$$\frac{dy}{dx} = C_1 (A - y)$$

Macy (1936) divided the entire response curve into three regions. This is approximatic Mitscherlics equation in three straight lines.

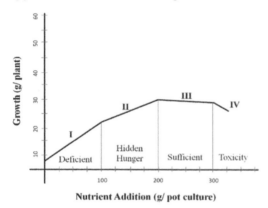

Fig 3. Response regions of crop to addition of nutrients.

I-region: Nutrients present in very small quantity causing deficiency symptom in plant. Greatest response to applied nutrients is observed.

II-region: Visible symptom of deficiency is not seen, but plant is suffering due to deficiency of nutrients called hidden hunger. There will be response to applied nutrients but not as large as in region-I.

III region: Sufficiency region- Further addition of nutrients will not lead to increase in yield in this region.

IV region: Toxicity region- Further addition of nutrients after sufficiency region will lead to toxicity.

Excess addition of iron, born and copper cause toxicity in plants, excess addition of nitrogen, phosphorus and potassium cause salt injury, excess of potassium cause imbalance of calcium and magnesium, excess of calcium cause iron deficiency (iron chlorosis), excess of nitrogen cause lodging in cereals, increase in pest and disease incidence.

Response to micronutrients generally does not follow that of macronutrients. It has distinct regions of deficiency, critical and toxic.

QUESTIONS

I. Answer the following

1. What are the factors restrict the availability of plant nutrients?
2. What is the relationship between soil reaction and plant nutrient availability?
3. Explain the antagonistic interaction of plant nutrients.
4. Explain the response of plants to single deficient nutrient according to Liebig?
5. Describe the Mitscherlich's concept of response of plants to nutrients.
6. Illustrate the Macy's (1936) concept of crop response to nutrients.
7. Illustrate the response pattern of crops to micronutrients.
8. State the physical factors restricting the availability of plant nutrients.
9. What are the competing factors of plant nutrients?

Chapter 3

Soil Organic Matter
Role, Dynamics and Management

Plants are the generators or the primary source of organic matter while animals are the transformers- secondary source of organic matter. In agriculture, plants add 10–30 percent of organic matter through their roots stubbles and leaf fall and exudations. The excreta of livestock, the organic wastes of urban and rural residential areas include the garbage and sewage and sludge. The by products of agro based industries *viz.* bagasse, vinasse, pressmud from sugar industries, the coir dust from the coir industries, dust of cotton from ginning industries, the fruit and food processing industries wastes.

BENEFICIAL ROLES OF SOIL ORGANIC MATTER

- Organic matter imparts brown or black colour to the soil.
- Soil organic matter binds individual soil particles into aggregates that are responsible for good structure and porosity distribution which allow aeration, root penetration and movement and retention of water. Polysaccharides are formed by the action of microbes which act as cementing agents which improves the aggregate stability.
- Humus is adsorbed on the surface of clay particles. The bonding of organic complex to clay surfaces can be achieved by cation bridges, hydrogen bonding, vander walls forces and sesquioxide humus complexes *etc*. In soils of high base status in fine sand and silt size, micro aggregates consists largely C-P- OM (clay polyvalent metal – organic matter) complex.
- Undecomposed organic matter applied on the soil surface reduces soil erosion, shades soil (prevents rapid moisture loss), soil organic matter act as insulators for movement of heat between atmosphere and soil. Organic matter keeps the soil cooler in warmer season and warmer in cooler season.
- Soil organic matter moderates the plasticity, cohesion, reduces crusting and bulk density of soils (1.60 g per cc).
- Soil organic matter improves the water holding capacity of soil, conserves soil moisture. The water holding capacity of soil with 5 per cent organic matter is 57 per cent while that of 3 per cent organic matter is 37 per cent.

- Soil organic matter reduces soil erosion, undecomposed soil organic matter left on the surface may reduce the impact of raindrops on soil particles, increases infiltration of water, increase in water stable aggregates, resist the erosion.
- Soil organic matter increases the cation exchange capacity due to presence of large number of functional groups present on the colloidal surface *viz.* carboxyl (Acid group), hydroxyl (Phenol or alcohol group), carboxyl (Keto or ester group), amide group (Ammonium group), sulphonyl group and phosphonyl group.
- Loss or gain of hydrogen leaves the negative or positive charge on humus. These negative charge cause cation exchange of humus. Cation exchange capacity of humus is 2–30 times higher than the mineral colloids on mass basis. The cation exchange capacity of organic colloids augments the nutrient holding capacity of the soil. Heavy metals (*viz.* lead, cadmium), herbicides and pesticides are held on the soil. Organic matter accounts for 20–90 per cent of the adsorbing power of the mineral soils. Organic matter is the source for buffering of soil.

Chemical

- Organic matter is the source of carbon for many microbes such as free dinitrogen fixers *viz. Azotobacter, Azospirillum*
- The carbon dioxide released during decomposition of organic matter reduces the soil pH due to formation of HCO_3.
- Soil organic matter accounts for 92–95 per cent nitrogen in soil. A furrow slice contains 0.02–0.40% N. Humus increases the availability of nitrogen.

The amount of organic nitrogen mineralized each year can be readily calculated if we know the mineralization rate, fraction of organic matter present, fraction of organic matter present as nitrogen.

Amount of organic matter mineralized = $2.24 \times 10^6 \, x \, y \, z$

Where x is the organic matter content of the soil

y is the nitrogen content of the organic matter

z is fraction of organic nitrogen mineralized per year

if the soil contains 5% organic matter, organic N content is 5%, where 2% mineralized per year.

Organic N mineralized = $4 \times 10^6 \times (5 \times 10^{-2}) \times (5 \times 10^{-2}) \times (2 \times 10^{-2})$

$\qquad\qquad\qquad = 4 \times 5 \times 5 \times 2$

$\qquad\qquad\qquad = 200$ kg N per ha in one year

The mean soil nitrogen used is 5% for surface soils weighing 4×10^6 kg or 10–25 cm soil layer which is much higher than tropical and subtropical soils. The proportion of mineralization depends on several factors related to soil properties which affect biological activity in the soil and nature of organic nitrogen compound present.

- Soil organic matter accounts for 50–60% phosphorus. Citrates, oxalates, tartarates released during the decomposition combine with iron and aluminium forming chelates. There by reducing the fixation of phosphorus. The proportion of organic phosphorus present is much less (less than half). Accurate measurement of organic phosphorus mineralized and made available to plants is difficult. The difficulty is associated with the rapidity with which any organic phosphorus released is adsorbed or precipitated. The organic phosphorus mineralization is much slower because of the presence of inositol phosphates which are stable. Organic phosphorus makes relatively a small contribution to plant nutrition except where fresh material is present.

- Sulphur contribution of soil organic matter is 80 per cent. Mineralization and contribution of organic sulphur to plant nutrition resembles that of organic nitrogen. Sulphur requirement of plants is one-tenth of nitrogen and sulphur content of soil organic matter is one-tenth of nitrogen content. Sulphur is slightly less readily lost from soil than nitrogen. Thus if mineralization of organic matter releases sufficient nitrogen for crop growth, the sulphur from organic matter should be normally be more than sufficient for plant needs.

- Large quantity of boron and molybdenum and other micronutrients are held in organic form.

- Soil organic matter plays significant role in release of micronutrients from minerals by the action of acid and humus.

- Soil organic matter chelates metal ions $viz.$ iron, zinc, copper and manganese. A chelate is an organic compound that can bond to a metal by more than one bond and form a ring or cyclic structure by that bonding. The solubility of chelates helps to mobilizes and increase the availability of these nutrients.

- Humus H^+ reacts with mineral bases extracts. These bases are loosely held on humic colloids and are easily available to plants.

$$H\ [Humus] + K\ [Al\ Si_3\ O_8] \rightarrow H\ [Al\ Si_3\ O_8] + K\ [Humus]$$

- Reduces the toxicity of aluminium and manganese in acidic soils.

Unidentified effects of soil organic matter

- Release of vitamins, hormones, aminoacids and other substances from newly dead roots and microbes in soil solution. Indole acetic acid, gibberilic acid and DIHB (3, 5 – di iodo -4- hydroxybenzoic acid) are the growth hormones released into the soil solution by soil organic matter.

- Increase the activity of enzymes like urease, cellulase, dehydrogenase and amylase which help in decomposition of organic matter.
- Reduces potassium fixation in vermiculites
- Improves the vitamin and aminoacid contents in the produce.
- Reduces nitrate pollution
- In low active clays (LAC) humus is the major contributor for negative charge.
- Some of the organic sources *viz.* straw and neem seed cake check the proliferation of plant parasitic nematodes. Humified organic material will reduce the toxic effects of pesticides and herbicides to crops.

Causes for SOM decline in soils

- Soil erosion
- Excessive tillage
- Burning crop residues
- Scarce use of organic manures and heavy dependence on fertilizers
- Continuous monocropping

Strategies to increase soil organic matter

Soil organic matter at 3–8% will improve plant growth. Under tropical climate it is costly to build up soil organic matter other than incorporating plant residues and disposing available organic manures into the soil. Soil organic matter is sustained by harmonious supply of growth factors for biomass production (*viz.* climatic, edaphic factors *etc.*) as well as nature of inorganic complexing matrices (mainly HAC and LAC) which are the main edaphic factors and major modifiers of soil organic matter. Addition of fertilizers increases the crop growth and root growth. Application of phosphorus and liming increases the growth of leguminous crops.

Tillage: Heavy tillage decreases humus content of soil by one-half from its virgin level. In steppe soils humus content is maintained at 15–20 cm. In Alfisols and Ultisols part of the humus migrated in complex with clay minerals to the *Argillic* horizon. In *Spodosols* it has migrated with oxides to the *spodic* horizon. In both these carbon sink is formed in the B horizon. Carbon content of soil is also lost due to repeated disturbance during cultivation.

Conservation tillage: Is any tillage sequence which reduces loss of soil and water relative to conventional tillage systems because of their effectiveness for controlling erosion, reducing run off and evaporation and retarding the rate of soil organic matter decline. The main features of conservation tillage are:

- Maintenance of soil surface roughness
- Maintenance of crop residues on the soil surface

Conservation tillage is achieved by non-inversion tillage or following only secondary tillage operations. The extreme case of conservation tillage is zero tillage. Planting and fertilizer application is done in one pass operation with minimum disturbance of the soil and surface residues. Conservation tillage promotes an increase in soil organic matter as a result of less tillage, greater return and fewer disturbances of residues and less erosion compared with conventional tillage. In semi-arid tropics the conservation tillage has greater scope as there is less biomass availability.

Mulching the surface soil with organic residues conserves soil organic matter besides adding organic matter to the soil. Largely available crop residues, leaf and litter from the forest area may be used as mulching material.

Cropping system in combination with conservation tillage has greater inf luence on soil organic matter. High residue producing crops in a no-till or reduced tillage system and legume or forage crops in rotation are especially more effective in slowing the decline of soil organic matter or increases the equilibrium level in the soil.

Grass and legume pasture- high net return of organic matter.

Cash or grain crop- low net return of organic matter.

Pasture based rotations will return more organic matter to the soil because of greater activity of soil fauna and flora, steady root growth and lack of soil disturbance.

Legume residues decompose fast hence there is faster depletion of organic matter. While cereal residues decompose slowly and there will be maintenance of soil organic matter. Organic matter with high C: N will be more effective in maintaining the soil organic matter.

Farming system approach will facilitate the recycling of the organic matter.

Alley cropping: Growing of alternate narrow strips of complimentary crops including deep rooted woody species. Strips of perennial vegetation, especially leguminous shrubs can add both fertility and organic carbon to the soil during the rotational period. Pruning of shrubs and trees during the growing season provides mulch or green manure for the cropped area.

Strip cropping: Crops of different height and canopy structure are grown in strips across the slope or direction of the wind in dry farming situation.

Ley farming: Is rotation of arable crops requiring annual cultivation and artificial pasture occupying field for two years or longer. Ley farming is difficult for adoption by small farmers. Large holders with livestock component can adopt ley rotation.

Cover or green manure crops: Pprotect the soil against erosion, increase infiltration and reduce runoff. Returning the biomass improve organic matter content of soil. The effectiveness of maintenance and accumulation of organic matter depends upon the composition of the crop, soil and climatic factors. Sunnhemp, cowpea, horsegram, *stylosanthus haemata* are ideal cover crops in dry lands.

On farm organic resources: Amount of crop residues available for return to soil is a key dominant factor influencing the equilibrium soil organic matter. Soil organic carbon is linearly related to the amount of maize stover returned to the soil over 11 year period. Higher grain yields indicate higher residue production and return to soil. Using combine harvesters return all the residues to soil. India has largest cattle population. The dung and litter generated should be diverted as manure.

Off farm organic resources

Urban wastes: Often contaminated with pathogens, undecomposable harmful substances *viz.* metal pieces, glass and heavy metals besides emitting foul smell. Urban wastes are stabilized by composting, which mitigates pathogen and foul smell.

Sewage wastes may be composted and used.

Agro industrial wastes: Press mud 4 to 5 million tons per annum is available in the country. Coir pith, fruit and vegetable wastes: India is the largest producer of fruits and second largest producer of vegetables.

Carbon- nitrogen ratio

When organic matter with wider C: N (> 40: 1) ratio are added to the soil the decomposing organisms viz. bacteria will grow and multiply at a slower rate as the nitrogen content in the added organic matter is insufficient to meet the growth and development. Hence they use soil available nitrogen and the nitrogen is locked up in the body of microorganisms which is temporarily not available to the plants. If crops are sown immediately after application of organic matter of wider C: N ratio they will suffer due to nitrogen deficiency. During this period general purpose microorganisms will be active while nitrifiers will be inactive. The period of non-availability of nitrogen to plants is called the period of nitrate depression which may last for few weeks depending upon the C: N of the material added.

Plant residues with C: N ratios of 20: 1 or narrower have sufficient nitrogen to supply for decomposing microorganisms and also to release for plant use. Residues with 20: 1 to 30: 1 supply sufficient nitrogen for decomposing microorganisms, but insufficient to meet the plant use while, residues with C:

N ratio of 30: 1 or more decompose slowly as they lack sufficient nitrogen for microbial growth and multiplication. Once the energy source is exhausted the general purpose decomposers will die. The nitrifiers become active. They act on dead microbial body and release nitrogen.

Table 3. C: N of the soil organic matter, microbial body and organic materials

Organic matter	C: N ratio
Soil organic matter	10: 1 to 12: 1
Microbial cells	4: 1 to 9: 1
Legumes and farmyard manure	20: 1to 30: 1
Soil humus	11: 1
Alfalfa	13: 1
Paddy straw or wheat straw	80: 1 to100: 1
Maize stalk	40: 1
Sawdust	225: 1

Carbon- nitrogen ratio is significant in management of crop residue and green manure. When crop residues with wider C: N is used the sowing of crop should be taken after 3 weeks of incorporation of crop residues. Delay in incorporation of green manure crops beyond reproductive stage may leads to increase in C: N ratio.

The rate of decomposition of the components of the organic matter will be in the order of rapidity is sugar, starch and simple proteins > crude protein > hemicellulose > cellulose> fats and waxes > lignins.

The rate of decomposition is faster under aerobic condition. The end products of decomposition are carbon dioxide ammonium, nitrate, sulphates and phosphates.

Under anaerobic decomposition the end products of decomposition are methane, hydrogen sulphide, organic acids, ethylene, carbon bisulphide *etc*.

The resistant end product of decomposition is humus. The extent of carbon dioxide evolved during decomposition of soil organic matter is to the extent of 25–30 kg per ha. The whole process of decomposition and release of inorganic forms of nutrients is called mineralization.

Various enzymes *viz*. cellulase, urease, phosphatase, sulfatase and protease produced by the decomposing organisms are involved in decomposition of organic matter.

Since the decomposition is a biological process environmental conditions *viz*. temperature, soil moisture, aeration, carbon – nitrogen ratio of the organic material, soil pH, presence of toxic materials *viz*. selenium, chlorine, manganese and aluminium and boron *etc*. determine the rate of decomposition.

Distribution of organic matter in the soil varies vertically and horizontally.

Table 4. Soil organic carbon content of different soils

Soil group	% OC
Deep black soils	0.3–0.8
Red and laterite soils	0.7–6.5
Alluvial soils	0.3–1.1
Hill and mountain soils	4.0–8.0
Desert soils	0.3–0.6
Coastal alluvial soils	0.5–0.9

Organic matter content is more in virgin soils. Coarse textured soils will have lower organic matter than clay soils. Poorly drained soils will have higher organic matter than well drained soils.

With depth the soil organic matter content will decrease. Deep plouging facilitate migration of organic matter to lower layers which is evident in formation of organic sink in argillic horizon in Alfisols and Ultisols, spodic horizon in Spodosols.

QUESTIONS

Answer the following

1. How does organic matter influence the physical properties of the soil?
2. Explain the role of soil organic matter in nutrient contribution and mobilization to plants.
3. What are the causes of soil organic matter decline?
4. Explain the strategies to increase soil organic matter.
5. What is the significance of C: N ratio of organic matter on nutrient release?

Chapter 4

Nitrogen: Role, Dynamics and Management

Greek scientist Theophrastus (300 BC) advocated legume rotation. As early as 1750 AD it was known that plants require Nitre. Louis Pasteure in 1860 AD stated the possibility of atmospheric nitrogen fixation by microorganisms. Hellriegel and Willfarth (1886) strengthened the views of Louis Pasteure by his observations on the irratic behaviour of legumes to nitrate nutrition. Beizerink (1890) isolated the organism involved in nitrogen fixation. The systematic field experiments conducted at Rothemstead experimental station showed that the plants respond to inorganic nitrogen.

ROLES OF NITROGEN IN CROPS

Plants contain 1-5 per cent nitrogen by weight. It is a structural component of chlorophyll, aminoacids and proteins. Plant proteins contain 16 per cent nitrogen ($6.25 \times \%N = $ protein). Plants are dependent on protein for their propagation. Cereals possess a threshold protein level below which grain will not form.

It is supportive to uptake of other nutrients.

It is essential for carbohydrate utilization. A certain amount of nitrogen must be present for optimum utilization of carbohydrates produced during photosynthesis. Under deficient condition excessive deposition of carbohydrates takes place in vegetative cells with consequent thickening of the cell wall, limited formation of protoplasm, reduced succulence and reduced growth. A growing plant must have continuous supply of free energy input for synthesis of macromolecules from simple precursors and for active transport of ions and other synthesizing materials throughout the plant. Carrier for this energy is adenosine triphosphate (ATP) another indispensable nitrogen containing compound.

It is a component of enzymes. Many enzymes associated with plant growth are all complex proteins containing nitrogen. Growth restrictions results from nitrogen shortages for production of molecules involved in the various enzyme systems.

Hormone molecules which are proteins/steroids or derivatives of aminoacids serve as coordinators of activity of different plant cells.

It stimulates the root development and activity. Total root mass as well as rooting depth is enhanced by optimal nitrogen availability. Extension of roots facilitates absorption of water and other nutrients required for growth. Classical examples include alleviation of drought effects where deep subsoil moisture is present. Uptake of fertilizer phosphorus is enhanced particularly when NH_4^+ has been applied. nitrogen application increases water use efficiency.

Nitrogen application increases the crop growth by increasing the leaf area. Hence plants are able to trap more light by increasing both the rate at which the cells multiply and thence their number. In cereals nitrogen increases the number of tillers/side shoots. Adequate nitrogen in plants along with phosphorus plays dominant role in maintenance of leaf area. Nitrogen increases the total dry matter and improves the yield parameters and crop yield. Excess of nitrogen causes self-shading. The lower leaves do not receive sufficient light. They turn pale green or yellow. They do not photosynthesize efficiently but continue to respire so that there is net loss of assimilates from the plant. Nitrogen responsive varieties should have oblique or acute leaf angle. If phosphorus, potassium and sulphur are deficient in soil, excess of nitrogen may delay the maturity of crops. Excess nitrogen induces copper and zinc deficiency in plants. In crops where the green leaves are the economic part of the plant improves the crop productivity *e.g.* cabbage, fodder, leafy vegetables. Nitrogen fertilization has favorable influences on the protein quality and bread making quality of wheat, milling quality of rice.

Adverse effect of excess nitrogen use on crops

- Excess nitrogen application in potato increases total drymatter but reduces the tuber yield.
- Excess nitrogen reduces sugar production in sugarbeet, oil content in oilseeds, but improves the protein content.
- Green leafy vegetables may accumulate nitrate nitrogen in the leaves.
- Formation of weak fibres in cotton and jute, synthesis of more proteins the plants become succulent.
- In cereals lodging may occur- due to increase in height of straw/weak stem due to inadequate accumulation of cellulose.
- In legumes nodulation is reduced.
- Quality of fruits is impaired.
- Quality of tobacco deteorates due to accumulation of more nicotine.
- Plants become excessively succulent and become susceptible to diseases *e.g.*, blast disease in paddy.

Nitrogen deficiency may occur due to:

- actual low level of nitrogen in the soil
- deficiency of molybdenum.

Under actual nitrogen deficient conditions plants will appear stunted, pale yellow. The yellowing appears first on the lower leaves. In severe cases leaves will turn brown and die. When nitrogen is deficient in plants the nitrogen compounds in the older leaves will undergo lysis. The protein nitrogen is converted into soluble form and translocated to active meristematic region. It is mobile element in plants. Nitrogen is the first element to become deficient in arid and semiarid regions of low rainfall areas due to poor addition of organic matter and its rapid decomposition, also in sandy soils in heavy rainfall areas lost by leaching and surface run off.

Nitrogen deficiency may develop in the presence of adequate levels of nitrate nitrogen in soils if the soils are deficient in molybdenum. Nitrate nitrogen must be reduced in the plant before it is utilized. Molybdenum is required for assimilation of nitrogen. If molybenum is deficient nitrate nitrogen accumulates in plant and at the same time plant may show nitrogen deficiency. In legumes nitrogen deficiency is observed in the absence of cobalt or inoculum. Cobalt is essential for symbiotic nitrogen fixation in legumes.

Forms of Nitrogen

Ninety five per cent of nitrogen in the surface soils usually occurs in organic form. The total nitrogen content of soils ranges from 0.02–2.5 per cent in sub soils and in peat soils. NH_4, NO_3 and NO_2 are the inorganic forms. The inorganic forms arise by the normal aerobic decomposition of soil organic matter alternatively through addition of fertilizers. These forms represent 2-5 per cent of soil.

Ammonium is the ideal source of fertilizer. When plants absorb ammonical nitrogen instead of NO_3 form one gram of assimilate yields 0.72g of amino acid on the other hand if NO_3 is absorbed 1 g of assimilate yields 0.49g of amino acids. Nitrates must be reduced before it can be incorporated into protein. This reduction process requires energy which uses two NADH molecules for each NO_3. Secondly persistence of NH_4^+ in soil may be the reason for its superiority over NO_3 as it is less prone to leaching. Normally plant uptake of ammonium form of nitrogen is more at neutral pH range.

Growth of plants is often improved when NO_3 and NH_4 are applied in combination rather than independently. Plants supplied with NH_4- N often contain lower concentrations of certain inorganic cations such as Ca^{++}, Mg^{++} and K^+ and higher concentrations of elements absorbed as anions such as SO_4^{--}, $H_2PO_4^-$, HPO_4^{--}, Cl^- as compared with tissues of plants receiving

NO_3^- nitrogen. Plants supplied with NH_4 nitrogen nutrition contain higher concentration of amino acids but lower accumulation of organic acids.

Nitrogen dynamics in soil

Nitrogen from organic matter is released through step by step process.

- Aminization
- Ammonification
- Nitrification

Aminization is brought about by heterotrophic microorganisms comprising bacteria and fungi. The products of aminization are amines and aminoacids. These are transformed into ammonium by the process called ammonification. The ammonium released may be absorbed by plants, may undergo immobilization, lost by volatilization or may undergo fixation. The ammonium fixation occurs in relatively expanding type of clay minerals *viz.* vermiculite, smectite and illite. Ammonium fixation is influenced by the factors like amount and type of clay mineral, freezing and thawing, wetting and drying, addition of organic matter presence of potassium and soil temperature. The fixed ammonium is in equilibrium with the exchangeable ammonium.

Only small portion of the fixed ammonium is available to crops, results in delayed uptake, and intensive cropping results in recovery of fixed ammonium. Fixed ammonium has residual effect. Ammonium fixation may be reduced by band placement. Application of potassium three weeks prior to application of ammonium reduces fixation. Wetting results in release of fixed ammonium. Oats has ability to use fixed ammonium.

Nitrification is the biological oxidation of ammonium to nitrate. It has two steps:

- Ammonia is converted to nitrite (NO_2)
- Nitrite to Nitrate

The first step is brought about by a group of obligate autotropic bacteria known as *Nitrosomonas.*

$$2NH_4^+ + 3O_2 \rightarrow 2NO_2^- + 2H_2O + 4H^+$$

The substrates from which the nitrite is produced include not only NH_4^+ but also amines, amides, hydroxylamines, oximes and a number of other reduced nitrogen compounds. Numerous heterotrophic organisms other than *Nitrosomonas* can reduce nitrogen compound to nitrite.

The conversion of nitrite to nitrate is affected by a second group of obligate autotrophic bacteria termed as *Nitrobacter.*

$$2NO_2^- + O_2 \rightarrow 2NO_3^-$$

Nitrosomonas and *Nitrobacter* together referred as *Nitrobacteria*. Nitrification results in acidification of soil, requires molecular oxygen *i.e.* aerobic condition.

Nitrification is influenced by supply of ammonium, population of nitrifying organisms, the rate of nitrite to nitrate formation is greater than nitrite formation in a well aerated soils. Soil reaction should be basic (8.5), adequate supply of calcium and phosphorus, soil aeration, soil moisture (80–90%) and optimum temperature (30–35 °C).

Denitrification is defined as the microbial reduction of nitrate or nitrite to gaseous nitrogen either as molecular nitrogen or as an oxide of nitrogen. It is a respiratory process present in a limited number of bacteria genera.

$$NO_3 \rightarrow NO_2 \rightarrow NO \rightarrow N_2O \rightarrow N_2$$

The bacteria which are potential denitrifiers are:

Alcaligenes, Agrobacterium, Azospirillum, Bacillus, Flavobacterium, Halobacterium, Hyphomicrobium, Paracoccus, Propionibacterium, Pseudomonas, Rhizobium, Thodopseudomonas, Thiobacillus, Chromobacterium, Corynebacterium.

Facultative aerobes involved in denitrification process are *Pseudomonas, Bacillus, Paracoccus.*

A few species of *Chromobacterium, Corynebacterium, Hypomicrobium,* and *Serratia.*

Several autotropes *viz. Thiobacillus denitrificans, Thiobacillus thioparus, Micrococcus denitrificans.*

The enzymes involved in denitrification are nitrite reductase, nitrate reductase A and B.

Denitrification is influenced by carbon (energy source) supply. Plant roots provide energy source through excretion of various organic excretions and deplete oxygen which are favourable for denitrification. In rice oxygen availability near the roots is more whereas deficient in soils away from the roots which favours synthesis of enzymes involved in denitrification. Generally oxygen level in soil is 4–17%. Submergence favours denitrification.

Low soil temperature slows denitrification process. At soil pH <5 nitrous oxide production is more while at >6 gaseous nitrogen is released.

Chemodenitrification is loss of nitrogen through intermediatery products of nitrification

$$NH_4^+ \rightarrow NO_2 \rightarrow NO \rightarrow N_2O \rightarrow N_2$$

Significance of Denitrification

Reduced fertilizer nitrogen use efficiency. About 30% of nitrogen is lost through denitrification. Nitrous oxide released through denitrification is a potential greenhouse gas. Depletes ozone layer. Nitrous oxide is 180 times efficient as greenhouse gas as compared to CO_2. and currently accounts for 5–10 per cent of global warming.

Nitrate leaching: nitrates lost by leaching may contaminate groundwater.

Ammonia volatilization is more in calcarious soils (soils rich in $CaCO_3$) and in moisture scarce soils.

$$NH_4 Y + CaCO_3 \rightarrow \leftarrow (NH_4)_2 CO_3 + CaY$$

$$NH_4CO_3 + H_2O \rightarrow 2 NH_3 \uparrow + H_2O + CO_2$$

Factors influencing ammonia volatilization:

Amount of ammonia fertilizers added, quantity of organic manures added.

Type of soil (more loss in coarse soils), soil aeration and moisture content, soil temperature and soil pH.

Nitrogen transformations in submerged soils

In submerged soils applied ammonical nitrogen is nitrified to Nitrate- N. it moves down to anaerobic zone where denitrification takes place. Loss of nitrogen may be 20–40 per cent. Crop recovery is as low as 40 per cent. Even under best management 10–20% nitrogen loss is prone to occur by denitrification. By deep placement of ammonium sulphate 49% is recovered, 26% is remained in the soil near roots or incorporated into microbial tissue. 25% lost by diffusion of NH_3 to aerobic layer which undergoes nitrification and then denitrification. Nitrification also occurs around the roots.

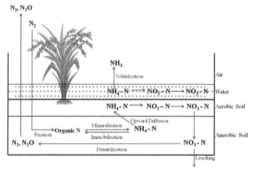

Fig 4. Nitrogen transformation in submerged soils

Nitrogen in submerged soils is lost due to ammonia volatilization when there is high concentration.

Nitrogen management in submerged soils

Placement of ammonical fertilizers in the reduced zone of submerged soils.

Maintain the flooded condition, alternate wetting and drying of paddy soil results in accelerated nitrogen loss.

Nitrification inhibitors are group of agrochemicals which block the conversion of ammonium to nitrate specifically inhibiting *Nitrosomonas* but not *Nitrobacter.* This would maintain fertilizer in NH_4 form for extended periods and provide for increased efficiency through enhanced uptake and yield. Unlike nitrate- N ammonical- N is less subject to leaching. Inhibition of nitrification will provide more opportunity to absorb ammonical nitrogen.

Desirable characters of nitrification inhibitors:

- Mobile in soil
- Persistent
- Nontoxic to non-target organisms

Situations where nitrification inhibitors are useful

In sandy soils, heavy rainfall areas, submerged paddy where leaching losses are high.

For crops requiring higher nitrogen e.g. Paddy, sugarcane maize, wheat.

Chemical nitrification inhibitors

1. Nitrapyrin or N- serve (NP) (2- Chloro – 6 – (trichloromethyl) pyridine) manufactured by DOW chemicals of USA in 1962 and Toyokoatsu Co of Japan. Effective at 1 ppm conc.
2. AM (2- amino – 4 – Chloro -6- methyl pyrimidine) AM manufactured by Mitsui Toatsu
3. DCD- Di-cyandiamide (AUF of Sumitomo chem. Ltd manufactured in 1981. mixed with urea at 4:1)
4. Guanyl thiourea – 1-amido-2-thiourea or ASU used in Japan
5. Terrazole – 5- ethoxy -3- trichloromethyl -1.2, 4-thiadiazole used in US
6. Sulfathiazole (ST) -2-sulfanilamidothiazole: used in Japan
7. TU- Thiourea
8. Dwell – Terrazole
9. MBT – 2 – mercaptobenzothiazole
10. Acetylene – C_2H_2
11. ATC- 4-amino -1,2,4 – triazole hydrochloride used in Japan
12. DCS – N- 2, 5- dichlorophenylsuccinic acid
13. Biocide
14. CaC_2 at 20 kg per ha along with urea drastically reduce denitrification nitrogen loss.

Organic nitrification inhibitors

Neemcake, pongamia (karanj) cake, castor oil, and acetone extract of seeds of neem, or pongamia, karanj bark and leaves inhibit nitrification. The active compound responsible for inhibition of nitrification is thought to be meliacins (epinimbin, nimbin, desacetyl nimbin, salanin, desacetyl salanin and azadirachtin).

Table 5. Nitrogen uptake and grain yield of wheat as influenced by the organic nitrification inhibitors

Treatment	Grain	Straw	N uptake
100kg N +44kg Neem cake	28.9	44.0	135.1
100kg N + 5% petroleum eather extracts of neem	23.5	35.6	91.5
100kg N +5% alcohol extract of neem	23.8	33.9	93.3
CD at 5%	0.5	1.3	-

(Indian J. of Agil. Sci. 49, 273-276)

Response to nitrification inhibitors

Nitrification inhibitors were found promising in rice, sugarcane, maize, potato and wheat.

In rice NP, AM, DCD increased the grain yields by 15–20%. Maize responded to nitrapyrin and Terrazole in coarse textured soils in which leaching losses are more while in clayey soils dinitrification losses are high.

In sugarcane the cane yields with nitrapyrin at 75 kg N per ha was on par with 150 kg N per ha. In addition there was significant residual effect of applied nitrogen. potato responded to nitrapyrin. There was significant saving in nitrogen (11–40 kg N ha^{-1}). Wheat yields were higher with 80 kg N per ha with DCD as compared to 120 kg N per ha without DCD. Yield increase with neem cake were also significant (4–12% increase)

Lipids associated with non-edible oils such as karanj oil, neem oil, castor oil, mahua oil imparts nitrification inhibitory properties. Nitrification inhibition was highest with castor oil (73.9) followed by karanj oil (64.2). The effectiveness of nitrification inhibition decreases with time. Waste mobile oil coating of urea granules also known to improve rice yields.

Table 6. Inhibition of nitrification by nonedible oils in wheat and rice

Treatments	Tillering		Panicle initiation	
	Wheat	Rice	Wheat	Rice
Neem oil	43.6	40.9	23.9	21.6
Karanj oil	64.2	60.8	57.5	52.8

Treatments	Tillering		Panicle initiation	
	Wheat	Rice	Wheat	Rice
Castor oil	73.9	58.5	49.6	45.6
Rattanjyoti oil	67.3	65.3	68.1	62.3
Mahua oil	49.1	45.5	35.4	33.6

(Nitrification inhibition as per cent of applied nitrogen; prilled urea applied at 50 kg per ha)

Urease inhibitors: Urea upon application to soil undergoes hydrolysis which is mediated by the enzyme urease. Urea hydrolysis may be minimized by inactivating urease and the substances used for this purpose are called urease inhibitors.

The mechanism of inhibition may be through inactivating the active sites on the enzyme (quinines, phenols and heterocyclic sulphur compounds). Structural analogues of urea which inhibit urease by competing with the same active site on the enzyme (*e.g.*, tiourea, methyl urea and substituted ureas). The compounds that react with the nickel atom in the urease enzyme (*e.g.*, hydroxamic acids).

A patented mixture of catachol and 2, 5 – dimethyl –P- benzoquinone added to urea completely inhibited urea hydrolysis for 10 days.

Essential characteristics of urease inhibitors.

They should be nontoxic, compatible with urea, move parallel to urea and effective at low concentrations.

Slowly available nitrogen sources

An ideal nitrogen source is one which releases nitrogen *vis-a-vis* crop demands throughout the crop growing period. Slow availability may be achieved by reducing the rate of solubility. Different coating materials are used to reduce dissolution, treated with materials to impede dissolution and granule size of the urea is modified.

Semipermeable membrane: The coating material may act as semipermeable membrane: Osmocoat used in high value crops. osmocoat 14 + 14 + 14 takes 3–4 months for complete dissolution while osmocoat 18 + 11 + 10 takes 8–9 months for dissolution.

Impermeable membrane with tiny pores: polyethylene films with limited number of very small pores used for fertilizer encapsulations.

Solid impermeable membrane: Substrates *viz.* gums, waxes, parafins, tars, asphalts, sulphur and a variety of materials are used. The coated material requires breakdown by microorganisms, chemical or mechanical process.

Sulphur coated urea (SCU): The nitrogen content varies from 36–38%. It is useful in sandy soils, in heavy rainfall areas and for long duration crops. In addition to nitrogen, sulphur is provided.

Tar coated urea (TCU): Tar or asphalt is coated to impede dissolution of urea.

Urea mud balls: Fertilizer encapsulated in the mudball. These mudballs are placed at the centre of the four hills of the rice. Fifty per cent recommended nitrogen may be economized.

Neem cake blended urea (NCBU): Prepared by using 1kg tar and ½ lit of kerosene to achive uniform coating. After coating the urea granules with tar, neem cake powder is added.

Urea also coated with gypsum called gypsum coated urea (GCU), mussorie phos is used as coating material called mussorie phos coated urea (MPCU), Urea blended with coal acid called as Urea Coal Acid, and humic acid is blended called Nitrohumic acid.

Urea treated with materials to impede dissolution

- Urea formaldehyde (30%)
- Crotonylidene urea (CDU) (30% N)
- Isobutylidene diurea (IBDU) (30%)
- Urea Z (UZ) (35%)

Urea granules modified to impede dissolution

The size of the urea granules is increased to slow down dissolution by reducing the area of contact for dissolution *e.g.*, large granular urea (LGU) and urea super granules (USG). Use of USG increased the grain yield of rice by 4–5q per ha over recommended urea.

Nitrogen use efficiency

Nitrogen use efficiency is defined as the unit quantity of grain (kg grain) per unit (kg) of applied nitrogen.

$$NUE = \frac{\text{Unit quantity of grain/economic produce}}{\text{Unit quantity of nitrogen applied}} \times 100$$

$$\text{Agronomic efficiency of nitrogen} = \frac{\text{Yield of crop with N–yield of crop without N}}{\text{Quantity of N applied}} \times 100$$

$$\text{Apparent recovery efficiency of } N = \frac{\text{N uptake by the N fertilized crop – N uptake in control}}{\text{Quantity of N applied}} \times 100$$

Nitrogen use efficiency in crop production rarely exceeds 40–50%. Large portion of the applied nitrogen is lost by volatilization and leaching. Mitigating nitrogen losses and effective utilization for crop production will overcome the adverse impact of nitrogen fertilizer use in agriculture on the environment. Increasing nitrogen use efficiency is also significant from the point of productivity and profitability of the crops. Since the fossil fuel is the source of major nitrogen fertilizers which are non-renewable. Efficient use of nitrogen fertilizers in crop production will extend the availability of scarce fossil based energy resources either naphta or coal.

Strategies to enhance the nitrogen use efficiency

Improving the ability of plants to compete with the process, which leads to losses of nutrients from soil- plant system to the environment - This can be achieved by genetical and biotechnological means. New plant types can be developed to use nutrients more efficiently so that there would not be much loss of applied nutrients.

Reduction in losses - which requires improved fertilizer use efficiency by improving soil and crop management practices/good agronomic practices.

Supplementing the fertilizer nitrogen with less expensive and locally available organic materials- organic manures, green manures, crop residues *etc.*

Integrating biological nitrogen fixation *viz. Rhizobium, Azospirillum, Azotobacter*, bluegreen algae, *Azolla, Acetobacter etc.*

Controlling the release of nitrogen by slowing nitrogen transformation process.

Good agronomic practices

These are basic for improvement of efficiency of fertilizer input.

Tillage: optimum tillage permits better root growth and proliferation. Conservation tillage reduces the loss of particulate organic matter. Fall plouging will improve the availability of nutrients from the previous crop residues. Deep tillage permit the nutrient exploitation from the deeper layers.

Selection of appropriate fertilizer responsive crop cultivars, timely sowing for full utilization to the growing season.

In the absence of optimum crop stand any agronomic practice fails to achieve higher yields. Hence optimum plant population should be maintained.

Irrigation water management and soil moisture conservation: often fertilizer use depends on the adequate soil moisture availability. Under scarce water situation irrigation water may be provided at the most critical stages for irrigation. Submergence in arable crops should be avoided. Alternate furrow or frequent shallow irrigation are most appropriate to mitigate denitrificaton losses.

Weed control: About 30–40 percent of the applied nitrogen is absorbed by weeds under unweeded condition. Adoption of integrated weed management mitigates nitrogen loss.

Cropping systems: Spacial and temporal combination of nonlegume and leguminous crops improve the nitrogen use efficiency by adding to the residual nitrogen effects. Exploration of nutrients from deeper soil layers due to the differential root patterns of the component crops and addition of leaf litter and crop residues improve the soil fertility.

Fertilizer nitrogen management

- Using optimum dose of nitrogen.
- Extensive fertilization under limited fertilizer resource.
- Selection of appropriate fertilizer material *viz.*
  Submerged paddy – ammonical fertilizers.
  Leguminous crops – diammonium phosphate.
  Combination of ammonical and nitrogenous fertilizer sources.
- Fertilizers should be placed in the reduced zone of the submerged soils. Deep placement in arable soils using appropriate farm implements *viz.* seed cum fertilizer drill.
- Timely application of fertilizers -Single large applications will increase the nitrogen concentration. Rapid losses due to volatilization and leaching of nitrogen occur. Nitrogen fertilizer application should match with the crop demand. In most of the cereals the peak uptake is from the beginning of seedling stage to beginning of the reproductive stage. Nutrient uptake follows the crop growth. Split application or fertigation will match the crop uptake and nitrogen availability in the soil. There must be network of roots in the soil for efficient nitrogen absorption. The nitrogen must be readily available in the soil for crop uptake.
- Foliar application and fertigation are efficient methods of supplying nutrients to the plants.
- Nitrogen use efficiency in rice increases with deep placement of ammonical fertilizers in the reduced zone of soil. Denitrification losses are reduced. Alternate wetting and drying accelerates nitrogen loss in flooded rice. Hence maintaining continuous flooded condition is essential.
- Balanced fertilization is the key for sustainable agricultural production and maintaining fertility of soil particularly where intensive agriculture is practiced. If any of the essential nutrients become inadequate, then it should

be timely supplemented in requisite quantity to maintain proper nutrient balance; otherwise, it may alter the availability of other nutrients and/or may restrict their absorption by plants leading to their poor growth.

- Nitrogen and sulphur applications increased the crop yields by 10 per cent in cabbage and fodder.
- Nitrogen and molybdenum applications are beneficial when nitrate is used as a source of nitrogen in root nodulating crops.
- Nitrogen and zinc improve protein, lipids, carbohydrate content of grain.
- Integrated nutrient management- improve the nitrogen use efficiency.

Nitrogen fertilizers

Anhydrous ammonia: It is a high analysis fertilizer (82% N). It is stored in pressure vessels at 1.725Mpa. It is applied to the soil by injection to a depth of 10 to 20 cm. Nitrogen loss by volatilization is high.

Aqua ammonia: (20–25% N): It is applied by injection method.

Ammonium nitrate: (33–34% N): It is used for top dressing. It is hygroscopic, risk of fire and explosions, less effective for flooded rice. Nitrogen is prone to leaching and denitrification.

Ammonium nitrate sulphate (30% N and 5–6% S): It is less hygroscopic.

Ammonium sulphate (21% N and 24% S): Less hygroscopic, stable, agronomically suitable. Not suitable for acidic soils. Low analysis fertilizer.

Ammonium chloride (25% N): It is more suitable for crops responding to chlorine *e.g.*, coconut and oil palms and soils having basic pH. It is largely used in Japan for rice.

Urea: Wholer a German scientist isolated urea from urine in 1773. Commercial production was first started in Germany in 1922. Urea is the largely used fertilizer in the world. Biuret present in urea is phytotoxic. Its concentration should not be more than 2%. Urea is hydrolysed to ammonia in presence of enzyme urease. Ammonia is further transformed into ammonium. High concentration of ammonia and ammonium are harmful to germinating seeds and seedlings.

Nitrate fertilizers

Sodium nitrate (16% N): It is naturally occurring along Chilian coast. Used in cotton and tobacco. Continuous use of sodium nitrate increases soil pH.

Potassium nitrate (13% N and 44% K_2O): It is less hygroscopic and has moderate salt index.

Calcium nitrate (15% N and 34% CaO): Known as Norwegian salt peter or lime nitrate. It is suitable for sodic soils. Used for foliar application in tomato, apple and celery. Extremely hygroscopic.

Potassium ammonium nitrate (16% N and 23% K): Known as kali ammonium salt peter.

Calcium Cyanamid (21% N) Known as nitrolime or lime nitrogen, does not absorb moisture, alkaline in reaction.

Table 7. Nitrogen fertilizers, nutrient content and equivalent acidity

Fertilizer	% N	%P$_2$O$_5$	% S	Equivalent acidity
Sodium nitrate	16			
Potassium nitrate	13		44	
Ammonium sulphate	21		24	110
Ammonium nitrate	33			
Ammonium chloride	25			128–140
Calcium nitrate	15			
Urea	45–46			80
Calcium cyanmid	22			
Anhydrous ammonia	82			148
Aqua ammonia	20–25			
Mono ammonium phosphate	11	48		
Diammonium phosphate	18	46		
Ammonium polyphosphate	12–15	60–62		
Calcium ammonium nitrate	25–26			Neutral

QUESTIONS

Define

1. Aminization
2. Ammonification
3. Nitrification inhibitors
4. Nitrification
5. Denitrification
6. Urease inhibitors
7. Agronomic efficiency of nitrogen
8. Nitrogen use efficiency

Answer the following

1. What is the role of nitrogen in crop plants?
2. What are the adverse effects of nitrogen in crop plants?

3. Illustrate the dynamics of nitrogen in submerged soils.
4. Describe the nitrogen management in submerged soils.
5. What are the mechanisms of slow release nitrogen fertilizers?
6. What are the modified urea granules? How they are superior to conventional urea?
7. Explain fertilizer management practices for enhancing nitrogen use efficiency.
8. Under what situation nitrification inhibitors are more useful?
9. Explain the crop response to nitrification inhibitors.
10. Why ammonical nitrogen fertilizers are advantageous than nitrate nitrogen fertilizers?

CHOOSE THE CORRECT ANSWER

1. The nitrogen content of ammonium sulphate is
   (a) 34 per cent   (b) 15 per cent
   (c) 20 per cent   (d) 26 per cent
2. The permitted limit of biuret content of urea
   (a) > 2 per cent   (b) < 2 per cent
   (c) < 0.2 per cent   (d) 3-5 per cent
3. Thiourea is a
   (a) Nitrification inhibitor   (b) Urease inhibitor
   (c) nitrogen fertilizer   (d) modified urea
4. Nitrite to nitrate conversion is brought out by
   (a) Nitrobacter   (b) Nitrosomonas
   (c) *Thiobacillus*   (d) *Pseudomonas*
5. Nitrate assimilation in plants is hindered if the soils are deficient in
   (a) Cobalt   (b) Calcium
   (c) Molybdenum   (d) Iron

Chapter 5

Phosphorus: Role, Dynamics and Management

Liebig (1840) on the basis of soil and plant analysis deduced the phosphorus requirement of plants. J.B. Lawes conducted series of field experiments to improve the fertilizer value of common sources of phosphorus such as bone meal, apatite, rock phosphate. Lawes (1843) produced super phosphate by treating apatite mineral containing phosphorus with sulphuric acid. He also stated that the phosphorus present in bones of higher animals can be made available by treating with sulphuric acid.

Phosphorus is the second most deficient nutrient in the world. Annual application of phosphorus fertilizers to crops is 36 million tons of which

1.32 m tones are consumed in India. Most of the arable soils in tropical and subtropical regions are deficient in phoshphorus, but the response of crops to phosphorus is uncertain. Phosphorus concentration in plants ranges from 0.1 to 0.4 per cent.

ROLE OF PHOSPHORUS

Phosphorus is the component of adenosine tri phosphate (ATP) and adenosine di phosphate (ADP) which are involved in storage and transfer of energy. ATP is involved in storage and transfer of energy. Transfer of energy from ATP to energy requiring molecule is called phosphorilation. Regeneration of ATP from ADP requires phosphorus at the sites of energy production.

Phosphorus is the structural component of nucleic acid, coenzymes, nucleotides, phosphoproteins, phospholipids and sugarphosphates. Adequate supply of phosphorus during the early stage of crop growth is required for laying down primordia for reproductive parts. Phosphorus is found in large quantities in growing parts, fruits and seeds. Phytin is the principal storage form of phosphorus in seeds. Phospholipid is another phosphorus containing compound forms the structure of protoplasm.

Phosphorus is associated with root growth and root proliferation. Adequate supply of phosphorus induces tolerance to root rot disease. Phosphorus induces early maturity of crops. Adequate phosphorus induces greater strength to cereal straw, improves the quality of fruits, forage, vegetables and grain crops. Adequate phosphorus induces tolerance to diseases. Winter damage or cold injury can be lowered by adequate supply of phosphorus in phosphorus deficient soils.

Deficiency of phosphorus: slow emergence and growth, off colour green foliage with purple venation especially on the underside of the leaves. Petioles have a purple cast, root development is poor. Decreased leaf growth, narrower leaves with tips green - violet or pinkish or greenish blue due to anthocyanin accumulation under severe deficiency, there will be drying of leaf tips and margins, decreased tillering, decreased photosynthetic activity and decreased translocation of photosynthates. In maize there will be excess accumulation of sugars. Bronzing and purpling are common in corn leaves. Purple colour venation in corn is associated with accumulation of sugars in phosphorus deficient plant. Biological nitrogen fixation is limited by phosphorus deficiency.

Phosphorus deficient plants look spindly and stunted. Older leaves are affected first. Low temperature may delay maturity and reduce tillering. Phosphorus application will induce winter hardiness.

Excess of phosphorus: Induce copper and zinc deficiency. The interaction of phosphorus with copper, zinc and iron are antagonistic.

Transformation of phosphorus

Forms of soil phosphorus

- Soil solution phosphorus
- Labile soil phosphorus
- Nonlabile soil phosphorus

Soil solution phosphorus present as primary ($H_2PO_4^-$) and secondary (HPO_4^-) orthophosphates. Plants largely absorb phosphorus in the form $H_2PO_4^-$, followed by HPO_4^-. The relation between soil solution phosphorus and pH is depicted in the figure.

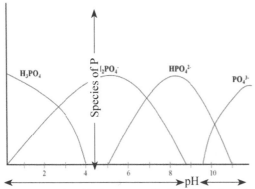

Fig 5. Relationship between soil pH and species of soil solution phosphorus.

Mechanisms of phosphorus uptake

Mass flow or convective flow: water absorption by plants which results in convective or mass flow of phosphorus. Mass flow would supply one per cent of the phosphorus needed by the plants in phosphorus unfertilized soil and 20 per cent in the fertilized soil.

Diffusion: phosphorus reaches absorbing roots by diffusion process. Major part of the plant requirement of phosphorus is met through diffusion. The factors responsible for diffusion are soil moisture, temperature, porosity, phosphate buffering capacity of the soil and rooting depth and root volume and root proliferation.

Phosphorus mobility in soil: phosphorus is less mobile in soil.

Maintenance of suitable concentration of phosphorus in the soil solution (phosphorus intensity depends on the solid phase phosphorus going into solution to replace the amount withdrawn by the plant uptake. The desirable soil solution phosphorus is 0.03 ppm. The quantity of solid phase phosphorus that acts as a reserve is frequently referred to as the capacity or quantity factor.

Soil solution P	$\rightleftarrows$	labile soil P	$\rightleftarrows$	Non labile P
Rapid		Slow reaction		
(H_2PO_4 and HPO_4)		(Fe, Al, Mn, Ca, Mg phosphate)		(Aged and well crystallized phosphates of Fe, Al, Mn, Ca, Mg and stable organic phosphates)

Labile phosphorus is the readily available portion of the quantity factor and it has high dissociation rate, permitting rapid replenishment of solution phosphorus. Depletion of labile soil phosphorus usually causes non-labile phosphorus to become labile again, but at a very slow rate. Thus the soil quantity factor is comprised of both labile and non-labile forms of phosphorus.

Phosphorus fixation/retention
Phosphorus is retained in the soil by three mechanisms

- Precipitation – dissolution reaction
- Sorption – desorption reaction
- Immobilization – mineralization reaction

Fixation is the collective term used to describe both sorption and precipitation reactions of phosphorus. The sequences of process *viz.* precipitation chemisorption and adsorption takes place continuously. Phosphorus forms less soluble compounds with Fe^{3+} and Al^{3+} at low pH, At neutrality forms soluble compounds Ca^{2+} and Mg^{2+}. Solubility of these compounds is highest between 6–7. Phosphorus retention in the soil is influenced by:

- Soil components *viz*. hydrous oxides of iron and aluminium
- Type of clay
- Calcium carbonate content
- Soil reaction
- Nature of ions either cations or anions
- Saturation of the sorption complex
- Soil organic matter
- Soil temperature
- Time of reaction

Influence of soil organic matter on phosphorus fixation
Soil organic matter reduces fixation of phosphorus by the following mechanisms.

- Humic acid released by the decomposition of soil organic matter forms phosphohumic complexes which are more easily assimilated by the plants.
- Humate anion of soil organic matter may replace the adsorbed phosphate ion.
- Formation of protective humus coating on the sesquioxide particle which prevents further fixation of phosphorus.
- By formation of clay –organic matter complex.
- Decomposition of easily decomposable soil organic matter results in evolution of carbon dioxide which forms carbonic acid which decomposes primary soil minerals and increasing phosphorus availability in neutral and alkaline soils.

Influence of soil reaction on phosphorus availability:
I. Strongly acid soils

$$Al^{3+} + H_2PO_4^- + H_2O \rightleftharpoons 2H^+ + Al\,(OH)_2\,H_2PO_4$$
$$Fe^{3+} + H_2PO_4^- + H_2O \rightleftharpoons 2H^+ + Fe\,(OH)_2\,H_2PO_4$$

II. Acidic soils

$$Al\,(OH)_3 + H_2PO_4^- + H_2O \rightleftharpoons Al\,(OH)_2\,H_2PO_4 + OH^-$$
$$Fe\,(OH)_3 + H_2PO_4^- + H_2O \rightleftharpoons Fe\,(OH)_2\,H_2PO_4 + OH^-$$

III. Moderately acidic soils

$$[Al] + H_2PO_4^- + H_2O \rightleftharpoons 2H^+ + Al\,(OH)_2\,H_2PO_4$$
$$[Al\,(OH)_2\,]OH^- + H_2PO_4^- \rightleftharpoons Al\,(OH)_2\,H_2PO_4 + OH^-$$

VI. Calcareous soils

$$Ca(H_2PO_4)_2 H_2O + 2CaCO_3 \rightarrow Ca_3(PO_4)_2 + 2H_2O + 2CO_2$$

Strategies for phosphorus management

The relationship between phosphorus availability and soil pH is illustrated in the figure.

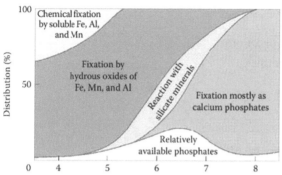

Fig 6. Relationship between soil pH and phosphorus availability

The total phosphorus is larger in the soil while phosphorus compounds of iron, aluminium, calcium and magnesiums is large whereas plant available phosphorus is small.

Solubility of phosphorus

Mono calcium phosphate: Ca $(H_2PO_4)_2$ is highly water soluble.

Dicalcium phosphate: $Ca_2 (HPO_4)_2$ is water and citrate soluble.

Tricalcium phosphate: $Ca_3 (PO_4)_2$ is insoluble in water and citric acid.

Citric acid is a weak organic acid. The root exudates contain organic acids which solubilize phosphorus insoluble in water which correlates citrate soluble phosphorus.

Phosphorus transformation in plant rhizosphere

Rhizosphere is the site of active phosphorus transformation. Rhizosphere has abundant bacterial and fungi (20-50 times more) population than in the bulk soil. Growing plants greatly influence the reactions in soil. Roots may absorb phosphorus from the soil or may exude into the soil. There may be competition for phosphorus between microorganisms and roots in the soil. Phosphorus availability in the rhizosphere is altered by the following mechanisms.

- Low molecular weight organic acids are secreted by or exuded from plant roots or microoganisms, and lower the rhizosphere pH or accumulate the dissolution of sparingly soluble phosphate minerals by complexing the metal cation of the mineral.

- Organic acid anions accumulate in sufficient concentration in the rhizosphere to compete effectively with orthophosphate for adsorption sites on iron or aluminium oxides.
- The plant alters the rhizosphere pH and hence modifies soil phosphorus solubility by net excretion of H^+ or HCO^-_3 to maintain a balance of electric charges associated with cations and anions crossing the root membrane.
- Soil phosphatases activity is increased; Acid phosphatase activity is higher in root rhizosphere, Alkaline phosphatase is released by microorganisms is also higher in the rhizosphere. Activity of both the phosphatases increases with the increase in age of the crop *e.g.*, gram, egyptian clover, wheat and mustard. The phosphatases render the soil organic phosphorus available to the crops.

Influence of mycorrhizae on the phosphorus availability to plants

Vesicular and arbuscular mycorrhizae (VAM) is an endomycorrhiza belonging to the family endogonaceae of phycomycetes. They possess two specialized structures known as vesicles (Bladder like structures) and arbuscles (Shrub like structures). The mycelium develops internally in the roots of growing crops and trees and produces arbuscles intracellularly. Vesicles are also produced by the internal mycelium but mostly intercellularly. These structures act as a storage organ inside the host plants root. Arbuscles act as the site of nutrient exchange. The hyphae enter the roots either through epidermis or root hairs and extend out as far as 2–8 cm from the root surface. The diameter of the hypha is 3–4 µm while that of root hairs is > 10 µm. Hence VAM is capable of mobilizing positionally inaccessible and less mobile nutrients *viz.* phosphorus, zinc, iron and copper to the plants. VAM has been found in cereals, grasses, legumes, cotton, potato, sugarcane; but exception in plants belonging to cruciferae and chenopodiaceae. VAM species *Glomus fasciulatum*, *Glomus mossae*, and *Glomus tenius* are used for pulses, oilseeds and vegetable crops. *Glomus microphylla* and *Glomus apigaeus* are suitable for crops grown in upland and medium land situation.

Mechanism of phosphorus mobilization by mycorrhizae

- Production of siderophores which chelate iron and release phosphorus.
- Presence of calcium oxalate crystals at soil-hyphae interface of mycorrhizal *Pascopyrum smithii* (Western wheat grass) which may be due to increased solubility of calcium apatite.
- Extra-radial hyphae of the fungal symbiont increase the absorbing surface of the root.
- Increase in the phosphatase activity observed in onion and wheat.

Phosphorus mobilization by mircroorganisms

- Microorganisms are capable of producing several organic acids *viz.* acetic, formic, lactic, keto gluconic, oxalic, succinic, malic and citric acids. These organic acids chelate calcium and iron. Such chelation results in release of phosphates into soil solution.

$$Ca_3PO_4 \rightarrow Ca_2(HPO_4)_2 \rightarrow Ca(H_2PO_4)_2$$

- Some of the efficient phosphorus solubilizing bacteria are *Agrobacterium radiobacter, Bacillus polymyxa* and *Bacillus megatherium.* Among the fungi *Aspergillus awamorii, Aspergillus flavous,* and *Aspergillus niger.*
- Microorganisms also produce inorganic acids. Sulphur oxidizing chemoautotropes produce sulphuric acid *e.g., Thiobacillus thioxidans, Pseudomonas striata, Thiobacillus ferrooxidans.*

$$S + 1\frac{1}{2}O_2 + H_2O \rightarrow H_2SO_4$$

The sulphuric acid so produced solubilizes insoluble phosphorus. Biosuper is produced by mixing rockphosphate with sulphur and inoculating with *Thiobacillus thiooxidans.*

- Carbon dioxide produced during respiration of microorganisms forms carbonic acid which also dissolves phosphorus.
- Humic acids and fulvic acids formed during degradation of organic matter in soil chelates iron, aluminum and calcium in complex phosphates releasing orthophosphates.
- Besides microorganisms produce acid and alkaline phosphatases which help in release of organically bound phosphorus.

Phosphorus transformations in submerged soils

- Phosphorus is released in submerged soil due to reduction of ferric compounds (free hydrous oxides of iron) with the liberation of sorbed and co-precipitated phosphorus.
- Higher solubility of $FePO_4 2H_2O$ and $Al PO_4 2H_2O$ resulted from hydrolysis due to increased soil pH in acidic and strongly acidic soils.
- Organic acids released by anaerobic decomposition of soil organic matter releases phosphorus by increasing the solubility of calcium bound phosphorus and complexing Ca^{2+}
- Mineralizaion of organic phosphorus
- Soil organic matter has promoting effect on reduction of ferric iron to ferrous iron. The ferrous iron is soluble.
- Release of phosphate anion by exchange between phosphate anion and organic anions from iron and aluminum compounds.

- Increase in the solubility of calcium bound phosphorus compounds in calcareous soils due to depression in soil pH as a consequence of carbon dioxide evolution during organic matter decomposition.
- Increased diffusion of phosphorus ions due to saturation.
- Increased phosphorous mobilization by the microorganisms due to presence of physiologically active rice roots which have the capacity to reoxidize the rhizosphere during the latter part of the growing period.

Phosphorus in pulses and oilseeds

Phosphorus requirement in grain legumes and oilseed crops is higher as compared to cereals. To produce one tone of grain legumes the crop absorbs 14 kg phosphorus, oilseeds absorbs 24 kg while cereals absorb 10 kg.

Phosphorus in grain legume seeds is stored as phytic acid. Phosphorus in combination with nitrogen and potassium improves the protein content of legumes while phosphorus in combination with sulphur, nitrogen and potassium improves oil content in oil seed crops. Phosphorus is the most limiting factor for production of pulses in drylands. Biological nitrogen fixation in legumes is hindered by phosphorus deficiency.

Pulses are largely grown in dry lands under moisture scarce situation. Phosphorus promotes root development and proliferation which increases water use efficiency and tolerance to drought.

Residual phosphorus

The recovery of the applied phosphorus is rather very low averaging < 20 per cent. Remaining 80 per cent of phosphorus accumulates in the soil. In calcareous soils it is found in the form of $Ca_8H_2(PO_4)_6\,5H_2O$. Phosphorus fertilizer application every year results in phosphorus build up in the soil. A stage may be reached where crops may not respond to applied phosphorus. Excess phosphorus build up in soils may result in loss through flood water and environmental concerns and nutrient imbalances leading to antagonistic interactions of nutrients in soil and plants.

Residual phosphorus is significant in high intensity cropping. Residual effect of phosphatic fertilizer depends on the

- Rate and frequency of application and duration of the crop
- Solubility of phosphatic fertilizers
- Soil properties
- Type of crops, yield levels, and extent of phosphorus removal

In general phosphorus accumulation in soil is affected by the rate of addition and removal in intensive cropping system. Four situations arise depending on the application rate and crop removal of phosphorus.

Low addition High removal Depletion of phosphorus	High addition High removal Maintenance of phosphorus
Low addition Low removal Maintenance of phosphorus	High addition Low Removal Build-up of phosphorus

- Rock phosphate application results in greater residual effect than completely acidulated rock phosphate.

Crop response to residual phosphorus

Rice can utilize iron bound phosphorus while legumes can use calcium bound phosphorus.

Wheat is a poor user of residual phosphorus as compared to rice. Hence in rice- wheat cropping sequence entire dose of recommended phosphorus for rice may be applied to the preceding wheat crop. Soyabean also responds to residual phosphorus of preceding wheat crop. Groundnut can utilize residual phosphorus applied to the preceding maize or groundnut crop. Ratoon sugarcane uses the residual phosphorus applied to the planted cane. Potato is the least efficient user of residual phosphorus because of its poorly developed root system.

Phosphorus requirement of crops and efficient use

Phosphorus fertilizer requirement is determined both by buffering capacity of soils and intensity of phosphorus in soil solution. Fixed phosphorus is the capacity factor. Efficiency of the added phosphorus increases several folds when the fixed phosphorus is present either in the organic or inorganic form. When phosphorus is added to soil with higher content of residual phosphorus the labile pool of available phosphorus increases beside the capability of the soil to absorb phosphorus is reduced. This results in more efficient use of phosphorus.

Phosphorus requirement varies with the crops and management practices. Fox et $al.,$ (1990) suggested the external and internal phosphorus requirement of crops. The external phosphorus requirement is the target phosphorus concentration in the soil solution that is associated with near maximum yield. On the other hand the internal phosphorus requirement is associated with maximum growth and yield. Phosphorus requirement by the crops can be calculated by a simple model proposed by Driessen (1986).

$$D = \frac{U_m - U_0}{R}$$

Where,

D = fertilizer phosphorus requirement.

U_m = phosphorus uptake by the crop at an optimum yield level.

U_0 = phosphorus uptake by the crop from unfertilized soil.

R = recovery of the applied fertilizer phosphorus (approximately 20 per cent)

Loss of phosphorus from soil

- Phosphorus is lost from the soil by erosion
- Removal by crop and weeds
- Leaching from the soil

Rationale of phosphorus fertilizer use

- Reduce the phosphorus fixation in the soil
- Phosphorus should be made accessible to plants
- Solubilize insoluble phosphorus present in the soil
- Select appropriate phosphorus fertilizer sources

Phosphorus fertilizer management strategies

Rate of application: At lower doses (40 kg P_2O_5 per ha) water soluble phosphorus fertilizers are superior in the early stages of the crop growth. In the later stages partly water soluble and citrate soluble phosphorus fertilizers are effective. At higher rates both water and citrate soluble phosphorus fertilizers are effective.

Crop duration: Short duration crops need water soluble phosphorus fertilizers. Combination of water soluble and citrate soluble phosphorus fertilizers are equally effective for medium duration crops in soils with high phosphorus retention and high phosphorus requirement. Long duration crops, perennials with extensive root system, permanent pastures and meadows need citrate soluble phosphorus fertilizers.

Soil phosphorus status: for soils having low phosphorus content phosphorus should be band placed. For the crops grown in soils which are low in phosphorus at least 20 kg P_2O_5 should be supplied through water soluble phosphorus fertilizer to meet the initial phosphorus requirement.

Soil reaction and phosphorus fertilizer: Granular phosphorus fertilizers with high degree of solubility are more effective than powdered fertilizers for acidic, neutral and calcareous soils. Granulation improves the effectiveness of the water soluble phosphorus fertilizers. Rock phosphate is a good choice for perennial crops grown in acid soils provided the soil contains some initial phosphorus, but for soils low in phosphorus partially acidulated phosphorus is ideal.

Placement: Band placement improves the effectiveness of the water soluble powdered phosphorus fertilizers in acid and neutral soils. Low water soluble phosphorus fertilizers (rock phosphate, bone meal) should be thoroughly incorporated into the soil. At lower doses band placement of water soluble phosphorus is preferred.

Yield levels: Regardless of crop or soil phosphorus levels for yield maximization water soluble phosphorus fertilizer is the ideal source.

Improving the effectiveness of phosphate rock

- Decreasing the particle size: eighty per cent of the ground particles should pass through 100 mesh (150 μm) sieve. This also increases the possibility of root interception.
- Mixing ground rockphosphate with water soluble phosphate fertilizers in 1:1 proportion.
- Phosphate rock is granulated with sulphur. Phosphorus availability is enhanced due to dissolution of insoluble phosphorus of rock phosphate by sulphuric acid which is produced by oxidation of sulphur. Inoculation of granulated phosphate rock - sulphur with bacteria *Thiobacillus* sp., fungi *Penicillium* spp. and *Aspergillus foetidus* enhance the phosphorus release from phosphate rock. Inoculation with *Aspergillus awamori* enhance water and citrate soluble phosphorus fraction of rock phosphate.

Phosphoric acid treatment of phosphate rock accounts for about 83 per cent raw material cost of triple super phosphate production. Sulphuric acid required for production of single super phosphate accounts for the major cost of the fertilizer. In view of economizing the cost of phosphorus fertilizer partial acidulation can be done.

- By direct acidulation of phosphate rock with less than the Stoichiometric amount of H_3PO_4 and H_2SO_4 required to make triple superphosphate or single superphosphate respectively.
- By adding rock phosphate to an immature single super phosphate reaction mixture and granulating the final mixture.
- Dry mixing of single super phosphate and triple superphosphate with unacidulated phosphate rock.

Fertilizer use efficiency is defined in the sense that plant nutrients are applied in such chemical and physical forms and at such times and placements that maximum yield is obtained with minimum possible amount of fertilizer nutrient application.

Why phosphorus use efficiency is low?

- Mobility of phosphorus is poor in the soil
- Concentration of phosphorus in soil at any time does not exceed 0.1 ppm
- Phosphorus ions are rapidly adsorbed by soil colloids or precipitated as iron, aluminium or calcium phosphates.

Phosphorus fertilizer management for enhancing the phosphorus use efficiency:

- Time of application: phosphorus requirement is more during early crop growth for promoting root growth and tiller production. Hence basal application is followed. In situations where initial soil status of phosphorus is high top dressing can be followed.
- Placement of phosphorus fertilizer: the mobility of the phosphorus is poor in the soil. Phosphorus diffusion is 0.04 cm in 10 days and roots in the soil are frequently 0.4 cm apart. In order to increase the utilization of applied phosphorus the plants either have more roots or nutrients need to be applied where the highest root density occurs. Placement reduces fixation of water soluble phosphorus and also increase the physical accessibility to the crops.
- Dipping the roots of rice seedling momentarily in single super phosphate mud slurry before transplanting economize fertilizer phosphorus by 50 per cent in highly phosphorus deficient soils as compared to conventional method of application at puddling. Dipping in diammonium phosphate slurry was also efficient.
- Foliar application of triple super phosphate may be followed.
- Seed soaking in phosphorus fertilizer solution increased the phosphorus uptake by the crop and crop yield. *e.g.*, dipping potato seed tubers in phosphate solution.
- Coating maize seeds with dicalcium phosphate was found more efficient than soil application.
- Increasing the moisture in soils which are low in phosphorus is likely to overcome phosphorus deficiency in crops.
- Mixed application of ammonium sulphate or potassium sulphate with phosphorus fertilizer enhanced the uptake of phosphorus by the crop.
- Use of phosphorus solubilizers and phosphorus mobilizers.
- Liming in acidic soils increases phosphorus availability.
- Blending citric acid soluble phosphorus fertilizers with organic manures or biogas slurry.

- Application of entire recommended phosphorus fertilizer to preceding green manure crop.
- Application of phosphorus fertilizer on the cropping system basis as a whole instead of individual crop basis.
- Phosphorus fertilizer application to the crops through fertigation

Interaction of phosphorus with zinc

Phosphorus can cause zinc deficiency because of increased crop growth and dilution effect. Phosphorus fertilization increases root to shoot ratio. As a consequence of decreased shoot growth due to zinc deficiency, phosphorus toxicity observed in plants even though total phosphorus content is not increased.

Zinc plays an important role in maintaining the integrity and selectivity of cell membranes. Any deficiency of zinc may result in loss of control on phosphorus uptake by the plants. Excess accumulation of phosphorus, nitrate, manganese, iron and copper was observed in zinc deficient soils.

PHOSPHATIC FERTILIZERS

Forms of phosphatic fertilizers

- Water soluble phosphorus: that portion of the fertilizer phosphorus soluble in water.
- Citrate soluble phosphorus: that portion of the residue soluble in 1N ammonium citrate.
- Citrate insoluble phosphorus: that portion in the residue remaining from water and citrate extractions is citrate insoluble phosphorus.
- Available phosphorus: The sum of water soluble and citrate soluble phosphorus represents an estimate of the fraction available to plants and is termed as available phosphorus.
- Total phosphorus: Available phosphorus + Citrate insoluble phosphorus

Phosphorus in the phosphorus fertilizers is expressed as P_2O_5. The relationship between phosphorus and P_2O_5 is:

$\% P = P_2O_5 \times 0.43$

$\% P_2O_5 = P \times 2.29$

Phosphoric acid (55% P_2O_5) and superphosphoric acid (76–80% P_2O_5): are used in preparation of liquid fertilizers, fertigation and for application through injection method in alkaline and calcarious soils.

Single super phosphate (16% P_2O_5 and 6–10% S in the form of calcium sulphate): It contains 90 per cent water soluble phosphorus. It is a low analysis fertilizer.

Triple super phosphate (44–52% P_2O_5 and <3% S): 95–98 per cent phosphorus is water soluble. It is a high analysis fertilizer.

Enriched superphosphates (25–30% P_2O_5 and variable in S): It is a high analysis phosphorus fertilizer. All the phosphorus present is available to the crops.

Monoammonium phosphate (11% N and 48% P_2O_5) and Diammonium phosphate (8% N and 46 P_2O_5) contain completely water soluble phosphorus. Used as starter fertilizers.

Ammonium polyphosphate (10–15% N and 35–62% P_2O_5): granular product

Rock phosphate (27–41% P_2O_5) the citrate solubility varies from 5–17% of the total phosphorus and none of the phosphorus is water soluble.

Rock phosphate with 12–17% citrate soluble phosphorus are rated as medium potential while those with less than 12% citrate soluble P are rated as low potential. Rock phosphate should be finely ground. At least 90% should pass through 100mesh screen (0.147 mm). It is applied at triple the recommended rate of water soluble phosphorus.

Basic slag or Thomas slag is the byproduct of the steel industry. It contains P_2O_5 and lime. It has neutralizing value of 60–80%. Hence it is useful in acid soils. During manufacture of high quality steel the impurities like silicon, sulphur and phosphorus combine with calcium and rise to the top of the furnace and are poured off. When solidified, it is finely ground such that 80% should pass through 100 mesh sieves and marketed as basic slag. Indian basic slag contains 2–7.24% P_2O_5 Some grades contain 17% P_2O_5 It also contains traces of zinc, copper, manganese and boron in.

Raw bone meal (20% P_2O_5): bones are crushed into finer particles of 2.24 mm. It contains 8% citrate soluble phosphorus.

Steamed bone meal (22% P_2O_5) is more effective. It contains 85% of the phosphorus which is soluble in 2% citric acid when ground to 150 to 200 mesh fineness. Available phosphorus is 16%.

Mussorie phos mined from the mussoorie region mountains in Dehradun in Himalayas.

Most of the phosphorus soluble in citric acid. It contains less fluorine. The mined rock phosphate is grinded so that at least 90% of the particles pass through 100 mesh.

Gapsa phos: imported from Tunisia. Mined in Gapsa in Tunisia, North Africa. Marketed in India by PPCL.

Table 8. Composition of Mussoorie phos and Gafsa phos

Composition	Mussoorie phos	Gafsaphos
P_2O_5	18 –20%	28–30%
Ca	37–45%	48.5%
MgO	4–7	0.8
S	4.1	
Zn	100–150 ppm	370 ppm
Cu	10–50 ppm	19 ppm
Mo	5–75 ppm	
Org. C	1–2%	1.0
Mn	0.18%	22 ppm

QUESTIONS

Answer the following

1. What is the role of phosphorus in crop plants?
2. Illustrate phosphorus dynamics in relation to soil pH
3. Explain the mechanism of phosphorus fixation in soils.
4. What are the management practices for efficient use of water soluble phosphatic fertilizers?
5. Explain crop and soil suitability for rock phosphate use.
6. Describe the management practice for rock phosphate use
7. How does soil organic matter enhance phosphorus availability to crops?
8. What is the mechanism of phosphorus mobilization by vesicular and arbuscular mycorrhiza to the crop plants?

Give reasons

1. Rock phosphate is incorporated into the soil while single super phosphate is band placed near the plant roots.
2. Phosphorus availability increases with submergence.
3. Zinc deficiency in soil causes phosphorus toxicity in plants.
4. *In situ* green manuring increases phosphorus use efficiency in crop plants.
5. Phosphorus application is recommended to the upland crops in rice based cropping system.
6. How do the microorganisms mobilize phosphorus in soil?

CHOOSE THE CORRECT ANSWER

1. The total phosphorus content of the steamed bone meal is
   (a) 15 per cent   (b) 22 per cent
   (c) < 12 per cent   (d) 26 per cent
2. An immobile nutrient in the soil
   (a) boron   (b) potassium
   (c) phosphorus   (d) nitrate
3. Energy currency in crop plants refers to
   (a) iron   (b) copper
   (c) potassium   (d) phosphorus
4. Single super phosphate contains
   (a) calcium   (b) sulphur
   (c) phosphorus   (d) phosphorus, calcium and sulphur

Chapter 6

Potassium: Role, Dynamics and Management

Glauber J. R. (1604–1670) in Netherlands first proposed that the salt peter (KNO_3) was the principle of vegetation. He obtained large increases in crop yield by addition of salt peter.

Mineral soils contain 0.04–3.0% K. Total potassium content of soils range from 3000 to 100000 kg per ha in the upper 0.2 m of soil profile. Of this total potassium approximately 98 percent is bound in mineral form whereas 2% in soil solution and exchangeable phases.

ROLE OF POTASSIUM

It is Involved in activating the enzymes (as many as 80) responsible for carbon dioxide reduction and required for synthesis of adenosine triphosphate. Hence deficiency of potassium reduces crop growth and yields.

Facilitates translocation of photosynthates, nitrogenous compounds from vegetative parts to the grain, thereby increases protein content of the grain.

Improves nitrogen fixation by *Rhizobia* in legumes with adequate potassium is ascribed to greater translocation of photosynthates from source to the nodules. Besides potassium activates the nitrogenase enzyme which is responsible for reduction of N_2 to NH_3 in the cells of the Rhizobium.

Adequate supply of potassium promotes formation of stronger supporting tissue which is the cause for reduced lodging in cereals. Improves the storability of fruits, imparts resistance to fungal diseases.

Adequate potassium supply results in incrustation of silica into the epidermal cell layers and increase in hemicellulose and lignin content in cell wall. Thus there will be formation of stronger epidermal layer and thickening of the cuticle. Besides the transpiration rate diminishes. As a consequence moisture content on the leaf surface decreases. Accumulation of free amino acids and sugars in the leaves also decrease. Thus creates unfavourable situation for fungal infection in plants supplied with adequate potassium. Reduction in the disease incidence was noticed for

• Leaf blight and stalk rot in corn

- Blast and stem rot in rice
- Wilt and damping of in cotton
- Mould and mildew in soyabeans
- Bunt and rust in wheat
- Blackspot and stem end rot in potato
- Late blight in potato
- Wild fire disease in tobacco
- Leaf spot and dollar spot in grasses.

Potassium reduces frost damage. Adequate supply of potassium to potato crop increases the potassium concentration in the leaves which in turn lowers the freezing point of cell sap. Thus the potato over comes frost damage.

Potassium is a quality element. Adequate supply of potassium fertilization improves the physical quality parameters *viz.* size, shape and colour of the kernel, tubers and fruits. It imparts strength, length and fineness of cotton fibres.

Biochemical quality parameters like protein, oil, vitamin C content, juice quality in sugarcane, storability of fruits.

Combustability of tobacco: Adequate supply of potassium supply in tobacco maintains optimum nicotine content, desirable K/Ca ratio and reduce total nitrogen in the leaves which impart good combustability of tobacco leaves.

In potato medium and large tuber yield increased with potassium supply which was attributed to increase in their number per square meter.

Adequate potassium fertilization reduces free amino acid and phenol content in freshly harvested tubers which is the desirable characteristic for processing.

In cotton ginning per cent, fibre length (mm) and bundle strength were increased with adequate potassium fertilization.

The per cent purity, sucrose and crushable cane sugar increased with increase in potassium fertilization.

Potassium is involved in the water relations. Provides much of the osmotic pull required for drawing water into the plants. Maintains turgor, which is essential for proper photosynthesis and metabolic process. Malfunctioning of the stomata is due to deficiency of potassium is correlated to lower photosynthesis and less efficient use of water. Potassium can affect the rate of transpiration and water uptake through regulation of stomatal opening. Improved potassium nutrition reduced the transpiration due to smaller stomatal aperture size in peas. Under adequate potassium supply osmotic potential of the plant water is greater than atmospheric water which may be the cause for reduced water loss.

TRANSFORMATION AND DYNAMICS OF POTASSIUM

Forms of potassium

- Soil solution potassium
- Exchangeable potassium
- Nonexchangeable potassium
- Mineral form of potassium

$$\text{Nonexchangeable K} \underset{}{\overset{\text{Slow}}{\rightleftharpoons}} \text{Exchangeable K} \underset{}{\overset{\text{rapid}}{\rightleftharpoons}} \text{Soil solution K}$$

Effectiveness of soil solution potassium is influenced by other cations *viz.* Ca^{++} and Mg^{++}. The availability potential of the soil solution potassium can be satisfactorily estimated by equilibrium activity ratio (AR_e^K)

$$AR_e^K = \frac{a_K}{\sqrt{a_{Ca+Mg}}}$$

This ratio is a measure of the intensity of labile potassium *i.e.* the potassium which is immediately available to the plants. AR_e^K of potassium indicates the momentary availability of potassium. Upon depletion of potassium by crops or by leaching the exchangeable potassium replenishes soil solution potassium. Two soils even if they have same AR_e^k differ in their potassium supplying capacity as depicted below.

Types of clay	AR_e^K	PBC	$-\Delta K°$
Kimmeridge clay	0.0247	92	2.27
Red clay	0.0249	27	0.67

The AR_e^k of both the soils is almost same. But potassium supplying capacity of Kimmeridge clay soil is greater due higher potential buffering capacity.

Potassium status of the soil not only depends on the current potential of potassium in the labile pool, but also the way in which the intensity depends on the quantity of labile potassium present. Exchangeable potassium is held in the planar (p) position, known as unspecific sites, edge and inner positions (e and i) known as specific sites. Potassium held on e and i positions is released on exhaustion of potassium held on p position.

Potassium status can be predicted by Q/I relationship of Beckett (1964). It is used to predict both immediate (I) and subsequent (Q) availability of potassium to crops.

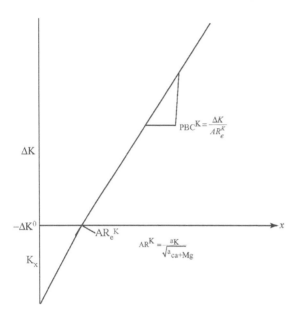

Fig 7. Q/I relationship for a specific Delvare soil

ΔK = quantity factor

AR^K = activity ratio for K or the intensity

ΔK^0 = Labile or exchangeable potassium

AR_e^K = equilibrium activity ratio for potassium

K_x = specific sites for potassium

PBC^K = potential buffering capacity

If $\Delta K > AR^K$ then PBC^K is more

   $\Delta K < AR^K$ then PBC^K is less

The AR_e^K value is a measure of availability of labile potassium in soil and it can be increased by potassium fertilization labile soil potassium may be more reliably estimated by ΔK^0 i.e., the size of the labile pool. Higher the values of $(-\Delta K^0)$ indicates greater potassium release into soil solution. Liming in soils having exchangeable Al^{3+} increases labile pool size as Ca^{2+} replaces Al^{3+} as the latter one is held at greater strength. Type of clay mineral influences size ΔK^0. Illite > Smectite > Kaolinite.

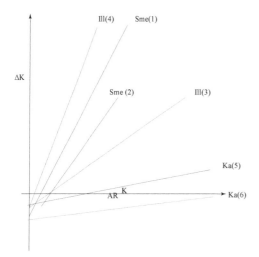

Fig. 8. Q/I relationship of soil of different dominating clay minerals and potassium saturation to 3% of CEC.

ΔK = quantity factor

AR^K = activity ratio for K or the intensity

Table 9. Clay minerology and CEC of the soils

S. No.	Clay minerology	Clay content	CEC of clay (meg/100g)
1.	Smectitic	44.7	86.0
2.	Smectitic	29.8	104.1
3.	Illitic	18.3	62.4
4.	Illitic	59.7	50.0
5.	Kaolinitic	54.9	17.8
6.	kaolinitic	70	11.8

Soils with kaolinite as dominant clay mineral contain only small quantities of labile potassium and have a low buffering (slope) for K^+. Soils with high activity clays like smectite and illite on the other hand, have a low K^+ concentration in their solution (AR_o), but larger potassium reserves and consequently higher buffer capacity (steep slopes).

Potassium availability in the soil solution

The PBC^K value is a measure of the ability of the soil to maintain the intensity of potassium in soil solution and is proportional to CEC. Soils having higher PBC^K signify a good potassium supplying power but a low PBC^K suggests need for frequent potassium fertilization. In sandy soils PBC^K is extremely small where CEC is mainly due to organic matter. Potassium is depleted in shorter time by leaching or by crop uptake. Such soils need frequent potassium fertilization. PBC^K is a constant provided non-exchangeable potassium is not altered significantly.

The ability of the soil to maintain a given activity ratio depends on the characteristics of the labile potassium pool, the rate of release of fixed potassium and diffusion and transport of potassium in the soil solution.

AR^K of potassium in soil solution is important since plants take potassium through diffusion. Even though certain soils *e.g.*, montmorillonitic soils of Texas (Nelson *et al.*, 1960) Q/I is high the value of I is too low to support plant growth. A great deal of potassium must be added to effect any significant change. In sandy soils even with less addition AR^K is maintained.

Drawbacks of Q/I concept

1. Measurement of potassium availability is laborious.
2. Invalid for soils having wide range of equilibrium concentration of calcium in soil solution
3. If Q/I curve is curvilinear then the interpretation is difficult.
4. Cannot be used to measure the amounts of potassium removed by exhaustive cropping of soil that yield initially non exchangeable potassium to plants.

Factors determining potassium availability to plants

- Kinds of clay minerals
- CEC of soil
- Potassium fixing capability
- Sub soil potassium and rooting depth
- Soil moisture
- Soil aeration
- Soil temperature
- Soil pH
- Calcium and magnesium content of soil
- Tillage
- Plant characters - root types, root CEC, rooting depth, root proliferation and density, genetic characters – variety or hybrid

Potassium uptake mechanism of plants

Nearly eighty five per cent of the potassium taken up by the roots is by diffusion mechanism. Plant root usually contact < 3 per cent of the soil volume in which they grow. Root volume accounts for < 1 per cent of the soil volume. (Barber 1985). The amount of potassium absorbed through root interception is negligible. There is a plant regulated reduction of concentration of potassium ions at the root surface. This may occur as a consequence of mass flow of potassium into roots resulting in concentration gradient in the soil solution. The steeper the gradient greater will be the diffusive flux. Rate of diffusion depends on soil moisture and concentration of potassium in the soil solution. Optimum supply of potassium to plants depends on sufficiently high level of potassium in the soil solution and adequate soil moisture and mobility of potassium in the soil. The rate of mobility of potassium ions in soil solution is 0.42 cm per minute (1 m in 4 hrs), whereas in dry soil mobility is negligible. Hence it is vital to maintain adequate potassium levels in soil to cause diffusive movement in the soil solution. The amount removed by the crop does not indicate the external demand by the crop.

Potassium in the rhizosphere

Plant roots deplete the soil solution potassium, thus concentration gradient is created between the root surface and the bulk soil that not only exchangeable potassium but also some interlayer potassium diffuses to the root surface. This phenomenon happens within the 1mm around the root surface *i.e.* in the zone of root hairs. In unmanured soils the depletion zone extends only 4.6 mm on the other hand with adequate potassium manured soils there was greater diffusion pressure gradient and the depletion zone extended upto 6.3 mm from the root surface. There was 3.5 times better potassium uptake of which 98 per cent came from the exchangeable pool of potassium.

Potassium fixation in soils

Potassium fixation occurs in soils rich in illitic, weathered micas, vermiculite, smectite and interstratified minerals. Potassium minerals are sufficiently small enough to gain entry into the silica sheets of the clay minerals where they become held very firmly by the electrostatic forces. Presence of NH_4^+ will alter the fixation of added potassium. Ammonium ions may trap the already fixed potassium. Potassium fixation is reduced in the presence of aluminium or aluminium hydroxy polymers. Potassium fixation depends on the concentration of potassium in the soil solution, wetting and drying, freezing and thawing. Upon drying potassium gets fixed. Freezing and thawing the soil helps in release of the fixed potassium. Fixed potassium cannot be estimated by using neutral normal ammonium acetate extractant. Fixation of potassium is not a total loss from the soil; it is a process of conservation of potassium from leaching and luxury consumption. Its availability is intermediate to exchangeable and non-exchangeable potassium.

Luxury consumption of potassium

It is the situation in which plants absorb excess of potassium than that is required for optimum growth. An element accumulates in the plant without a corresponding increase in growth or yield. It leads to inefficient or uneconomical use of an element.

Priming effect of potassium on nitrogen

Nitrogen uptake was significantly increased whenever nitrogen and potassium were applied together. This was attributed to the priming effect of potassium on nitrogen uptake.

Loss of potassium from the soil

Crop removal: Certain plants absorb potassium more than their requirement *e.g.,* perennial grasses (3.2–3.5% K), maize (1.7–2.1% K). When the residues of such crops are not returned to the same field there will be complete loss of potassium from the soil. Similarly in tea plantations the vegetative part is removed from the field.

Leaching loss: Applied potassium moves a short distance if the soil is predominant in potassium fixing clay minerals like vermiculite. Soils predominant in kavolinite type of clay retain less potassium. Coarse textured soils promote leaching of potassium. In high rainfall areas potassium is lost from the peat and muck soils as the bonding strength of potassium is less than other cations. Loss of potassium is high in soils with low CEC. Potassium leaching is enhanced if associated anion is chlorine than sulphate.

Table 10. Potassium removal by the crops.

Crops	Yield (M t ha^{-1})	Potassium removal (kg ha^{-1})
Wheat	4	83.6
Rice	5	89
Sugarcane	100	150
Groundnut	1.5	35.5
Cotton	1.8	232
Potato	10	92.6
Tobacco	2.5	57.5
Coconut	6.7	232

Soil erosion: Potassium is lost from the field through erosion in high rainfall areas with steep topography, where as in irrigated areas due to miss management of irrigation water *viz.* excessive flooding.

Irrigation with poor quality water: Use of irrigation water rich in sodium, calcium or magnesium results in displacement of exchangeable potassium from the exchange complex. Further it may be lost through leaching.

Strategies for potassium management in crop production

Judicious potassium management aims at minimizing the leaching losses, luxury consumption, potassium fixation in soils and utilizing natural potassium sources and recycling potassium through crop residues.

Leaching losses are high in soils with low CEC, sandy soils, soils having low pH (4–6) in heavy rainfall areas. In such situations split or frequent application of potassium fertilizers minimizes potassium loss. Liming of acid soils to maintain pH at 6.0–6.5 increases potassium adsorption on clay colloids, as a consequence potassium loss by leaching is reduced.

Timing of potassium application should coincide with peak demand of crops. The critical period seems to be during the early stages *i.e.* at the time of leaf expansion and tillering. During later crop growth stages potassium is lost from the older leaves by leaching. In barley 60 per cent of potassium is lost from the older leaves. In annual crops in low potassium retentive soils potassium application in 2–3 splits at maximum tillering and heading stages, while 2–3 times a year in perennial crops is beneficial. Split application also minimizes potassium fixation and luxury consumption. Leaching losses are uncommon in arid regions and in black clay soils.

Method of application: Localized or band placement in soils having low CEC increases leaching loss. Broadcasting part of potassium requirement or split application increases efficiency. Potassium being relatively less mobile in most of the soil, band placement is beneficial in soils low in potassium, high in potassium fixation capacity. Band placement increases potassium diffusion rate. If soil initial potassium status is high band placement has no advantage over broadcasting. High potassium levels in soil injure the seedlings. So placement of potassium should be at least five centimeter away from the seed row and to a depth of 2.5 cm. In dry lands band placement in moist soil increases utilization as it becomes unavailable if placed on the surface soil due to its rapid desiccation. Deep placement encourages deep rooting of the crops in contrast to shallow rooting with surface application. Broadcasting of potassium fertilizers is more effective in crops with shallow and active roots in soils rich in potassium.

Returning crop residues: Removal of crop residues depletes large quantity of potassium from the soil. Trash of sugarcane, stalks of sunflower and cotton, straw of wheat and paddy, stalk of maize, pseudostem and leaves of banana are good sources of potassium. They should be returned to the field. Growing deep rooted greenmanure crops *viz.* sunnhemp, dhaincha mines potassium from deeper soil layers and makes available to the crops upon incorporation to the soil. Green leaf manuring adds potassium to the soil from outside.

Potassium sources: Muriate of potash is satisfactorily applied for many crops. But for high value crops where quality is of prime importance sulphate of potash or potassium nitrate are preferred. Muriate of potash use in tobacco spoils the burning quality of tobacco. Use of sulphate of potash in potato increases storability of potato tubers and also specific gravity. Potassium nitrate has the advantage of low salt content.

Potassium Fertilizers

The potassium content of the potassium fertilizers is expressed as K_2O. The relationship of per cent potassium to K_2O is as follows.

$\% K = \% K_2O \times 0.83$

$\% K_2O = \% K \times 1.2$

Table 11. Potassium fertilizers

Potassium sources	Potassium content (% K_2O)	Features
Potassium chloride	60	Muritate of potash, derived from muratic acid. Largely used fertilizer
Potassium sulphate	50	Contains 17% sulphur, suitable for potato and tobacco
Potassium magnesium sulphate	22	Contains 11% Mg and 22% S,
Potassium nitrate	44	Contains 13% N, excellent fertilizer, used in cotton, tobacco and vegetables.
Potassium meta phosphate	38	It has 55–57% P_2O_5, it does not leave salt effect.
Potsssium polyphosphate	26	It contains 26% P_2O_5, it is liquid fertilizer
Orthopotassium phosphates	32	It has 52% P_2O_5, It is used for foliar application and for preparation of liquid fertilizers.
Potassium magnesium carbonate	24–27	Used in tobacco and fruit crops
Potassium carbonate	63–66	Used in tobacco

On farm available potassium sources

Wood ashes (4% K_2O)

Ash of lime kilns, brick kilns

Cotton hull ash (25% K_2O)

Fly ash: Potassium in fly ash is less soluble, calcium and magnesium may affect potassium uptake, contains boron at toxic level.

Furnace dust:

Kainite (12–14% K_2O): Untreated product from the mines.

QUESTIONS

Define

1. Potassium fixation in soil
2. Luxury consumption of potassium
3. Equilibrium activity ratio of potassium
4. Priming effect of potassium on nitrogen

Answer the following

1. State the role of potassium in crop plants.
2. What is the mechanism of disease resistance imparted by potassium in plants?
3. Why the potassium is called quality element?
4. Explain the relationship among different forms soil potassium.
5. Explain the mechanism of potassium uptake by the plants
6. How do secondary clay minerals of the soil influence potassium management for crops?
7. What are the types of potassium losses from the soil?
8. What are the potassium management strategies?
9. Why potassium use efficiency is more than nitrogen and phosphorus?

CHOOSE THE CORRECT ANSWER

1. The potassium fertilizer which spoils the quality of tobacco
   (a) Potassium carbonate   (b) sulphate of potash
   (c) potassium nitrate   (d) muriate of potash
2. The potassium availability in acid soils increases with the application of
   (a) nitrogen   (b) lime
   (c) phosphorus   (d) sulphur
3. Sodium induced potassium deficiency in sodic soils is alleviated by
   (a) potassium fertilizer application
   (b) management practices to reduce sodium content of soil
   (c) application of lime
   (d) addition of crop residues

4. The quantity intensity relationship (Q/I) of potassium was proposed by
   (a) Beckett   (b) Liebig
   (c) Glauber   (d) Lawes

5. 150 kg of muriate of potash contains
   (a) 75 kg K   (b) 90 kg K
   (c) 75 kg K_2O (d) 38 kg K

Chapter 7

Calcium, Magnesium and Sulphur: Role, Dynamics and Management

CALCIUM

Calcium is important in imparting structure and permeability to cell membrane. Deficiency produces general breakdown of membrane structure. Calcium enhances the uptake of NO_3^- N and interrelated with NO_3^- metabolism. It regulates cation uptake, required for cell elongation and cell division. Deficiency causes failure of terminal buds and apical tips of roots to develop. Blossom end rot of tomato, bitter pit of apples are due to calcium deficiency. Calcium is beneficial in imparting good soil structure.

It is immobile in plant. Very little translocation of calcium in the phloem is observed. Hence there is poor supply of calcium to fruits and storage organs. Downward translocation of calcium is also limited. Hence roots fail to enter the low calcium soils.

Calcium deficient soils

In acid and highly leached soils of humid areas calcium is deficient. Alkali soils, and some saline soils contain higher amounts of sodium are deficient in calcium. The calcium concentration in the root zone should be higher than the required by the plants.

Available calcium
- Soil solution calcium
- Exchangeable calcium

Calcium absorption
Calcium is absorbed in Ca^{2+} form. The capacity of the plants to absorb calcium is limited because it can be absorbed only by young root tips in which the cell walls of the endodermis are still unsuberized. The mechanism of absorption is by root interception or by contact exchange or by mass flow. Poor root growth reduce calcium uptake.

Calcium is generally lost by leaching. In arid region secondary deposition of calcium in the form of calcium sulphate or calcium carbonate is observed. Calcium in soil solution is in dynamic equilibrium with that on the exchange complex.

Soils developed from calcium bearing minerals *viz.* calcite, dolomite and gypsum are rich in calcium in arid regions. In humid regions soils are low in calcium due to leaching and precipitation of calcium. Calcium is deficient in low pH soils.

Calcium availability to plants is determined by the total calcium present in the soil, soil pH, CEC, per cent saturation on the clay colloids and nature of the complimentary ion.

Soils having high montmorillonitic clay require higher calcium saturation (70% or more) while kaolinitic clays require lower (40–50% Ca) calcium saturation to release sufficient calcium from the exchange complex for the growing plants. Under normal conditions the exchangeable calcium should be 60–85% of the total exchangeable capacity.

Calcium uptake is reduced if the complementary ions are NH_4^+, K^+, Mg^{2+}, Mn^{2+} and Al^{3+} while plants supplied with NO_3^-N stimulate the uptake of Ca. High concentration of calcium reduces the toxicity of Al^{3+} in acid soils and Na^+ in alkali soils.

Calcium and crop response

Calcium saturation on the exchange complex is directly associated with crop yields. Cotton yields are reduced if calcium saturation is below 40-60% and aluminium saturation is 40–60%. Soybeans are reported to suffer due to calcium deficiency at calcium saturations of 20% or less and aluminum saturations of 68% or more.

Calcium Fertilizers

- Single super phosphate (18–21% Ca)
- Triple super phosphate – (12–14% Ca)
- Calcium nitrate (19% Ca)
- Liming materials–Calcite, dolomite, hydrated lime, precipitated lime, blast furnace slag and gypsum (23.6% Ca)

Fineness of the liming material

Quantity required is inversely proportional to the fineness of the liming material. Limestone with at least 50% of the material passing through 60 mesh screen (Particle size of 0.25 mm diameter or less) are quite satisfactory for most of the agricultural purpose.

MAGNESIUM

Magnesium is the structural component of chlorophyll. Chlorophyll alone accounts for 15–20% of the total magnesium in plants. It is a component of ribosomes which are involved in the synthesis of protein. Hence magnesium deficiency reduces protein synthesis and non-protein nitrogen generally increases in the plants. It is involved in the activity of every phosphorilating enzyme in carbohydrate metabolism. Magnesium is involved in energy transfer and metabolic process in the plants.

Deficiency of magnesium in plants

Magnesium deficiency causes interveinal chlorosis of the leaf, wherein only veins remain green. In cotton lower leaves turn reddish- purple cast, gradually turning brown and finally necrotic. Premature leaf shedding is common. Maize, potato and cucurbits are the indicator plants of magnesium deficiency.

Grass tetany: Cattle consuming magnesium deficient forage crops may suffer from Hypomagnesemia or grass tetany. High rates of ammonical or potassium fertilizer use induce grass tetany. The optimum magnesium concentration in mature leaves is 0.20–0.25 per cent.

Magnesium is absorbed by the plants by mass flow or root interception. Magnesium concentration of 24 ppm is the approximate concentration satisfactory for most of the crops.

Magnesium is lost from the soil by leaching. Magnesium is adsorbed on the clay mineral or precipitated as secondary clay mineral.

Magnesium is deficient in acid sandy soils of high rainfall areas. Use of irrigation water having high Ca: Mg, high potassium use leads to magnesium deficiency. Nitrogen and phosphorus application along with magnesium increases magnesium concentration in the leaves. Excess of calcium decreases magnesium uptake. High amount of NH_4^+ -N, potassium and sulphate aggravates the magnesium deficiency. Calcareous soils, soils with low cation exchange capacity, Ca: Mg if greater than 7:1 induces magnesium deficiency. Satisfactory Ca: Mg is < 7:1.

Similarly K: Mg antagonism is of major concern in soils low in magnesium. In United Kingdom the satisfactory ratio of exchangeable K: Mg are developed. For field crops it is 5:1 while for vegetables it 3:1.

Magnesium fertilizers

Magnesium deficiency is corrected by soil or foliar application of epsom salt ($MgSO_4$ contains 9.8% Mg). Magnesium nitrate (16% Mg) or magnesium chloride (8–9% Mg) are more effective for foliar application than magnesium sulphate. Magnesia (MgO contains 55% Mg), dolamitic lime stone, $Mg(OH)_4$, Potassium magnesium sulphate (11% Mg and 22% K_2O), Magnesium silicate (basic slag contains 3–4% Mg) and serpentine (26% Mg) are also used to supply magnesium to crops.

Maize, tobacco, potato, cotton, sugar beet, and citrus are highly responsive to magnesium application.

SULPHUR

Horstmann (1911) established the essentiality of sulphur for plants. Sulphur is regarded as the fourth major nutrient. Its concentration in plants range from 0.1 to 0.5% (0.2 to 0.4% is optimum). It is a constituent of methionine (21% S) and Cystine (27% S). Sulphur is present in the disulphide bonds in polypeptides and proteins. Deficiency of sulphur inhibits protein synthesis, which is of special significance in leguminous plants. Sulphur is required for synthesis of co-enzyme 'A' and chlorophyll. It is vital constituent of ferrodoxins which, participates in photosynthesis. The characteristic properties of enzymes involved in photosynthesis and nitrogen fixation in nodules is attributable to type of the sulphur linkage present in these protein enzymes.

Influence of sulphur on the quality of crops

Sulphur application increases the oil content of oil seeds. Total oil yield increase with sulphur application was attributed to both increase in oil content of seeds and seed yield.

Application of sulphur and phosphorus markedly reduced the hydrocyanic acid (HCN) content in the fodder sorghum to the safe level of ruminant consumption.

The proportion of sulphur rich and sulphur poor proteins altered with sulphur fertilization.

N: S ratio and aminoacid composition altered, but the total concentration of proteins remained unchanged. Sulphur plays role in synthesis of storage proteins than metabolic proteins indicating its role in nutritional quality.

Sulphur deficiency reduces methionine even though methionine is non-essential it is used in the synthesis of cysteine. Hence both are important. Adequate sulphur fertilization of crops is essential to promote methionine synthesis. In soyabean sulphur fertilization progressively increased the cystine, cysteine, methionine, crude protein and oil content.

The N: S for efficient utilization of crude protein by the ruminants should be 10:1–15:1. Maize fertilized with adequate sulphur is preferred for poultry feed.

Wheat flour of adequate sulphur fertilized crop improves the baking qualities such as dough extensibility and resistance, and increase in loaf volume. Sulphur is a component of sulfhydryl and disulphide bonds of wheat gluten proteins. It increases cysteine and methionine content of wheat.

Sulphur fertilization increased methionine content in finger millet which will improve the ball making quality of finger millet flour. Sulphur along with the nitrogen improved the oil and protein content in groundnut, oil per cent and oil yield in rapeseed and mustard. Sulphur is a constituent of volatile acids found in the members of cruciferae and lilliaceae, imparts good flavour to mustard. Sulphur fertilization improved linolenic acid content of linseed. Sulphur fertilization favours transformation of oleic acid to linoleic acid and linoleic acid to linolenic acid. In forages if N: S is wider than 20:1 ruminants may not utilize the forages efficiently. Optimum N:S ratio is 10:1–12:1. Sulphur fertilization narrows down N:S ratio. In alfalfa Vitamin-A content increased with sulphur fertilization.

Sulphur deficient plants are chlorotic, with poor photosynthetic activity, formation of carbohydrates, proteins and oil is depressed. Rate of growth reduced; look spindly with short slender stems, reduced tillering in cereals, and poor nodulation in leguminous plants. Yield and quality of the crops are affected in less severe cases.

Sulphur deficiency is found in the younger leaves which turn yellow with lighter green veins. This is in contrast to nitrogen deficiency where protein nitrogen is hydrolysed in the older leaves and moves to younger/growing part. But sulphur compounds are stable. Organic sulphur (SO_4-S) found in vascular tissue, hence veins remain light green.

Sulphur toxicity is called *akagre* or *akiochi* common in ill-drained soils. It is characterized by inhibited root development and by browning and death of roots due to accumulation of H_2S on roots. This can be alleviated by draining water or by addition of iron fertilizer.

$$Fe + H_2S \rightarrow FeS \rightarrow FeS_2$$

Causes for sulphur deficiency

- Restriction of balanced use of fertilizer to NPK alone.
- Depletion of sulphur reserves due to intensive cropping, leaching and soil erosion.
- Non return of crop residues, scarce use of organic manures and depletion of soil organic matter.
- Use of sulphur free fertilizers, pesticides and fungicides.
- Greater restriction of industrial emission of sulphur.

Sulphur deficient soils

- Light textured soils which are low in organic matter.
- Regions of high rainfall.
- Soils irrigated with water low in sulphur.
- Continuous submergence, intensive cultivation of rice.

Forms and dynamics of sulphur in soil

Plants absorb sulphur in SO_4^- form. Sulphur is present in organic form, SO_4^- form and adsorbed sulphur and co-precipitated with calcium carbonate which is available to the plants.

Under anaerobic condition SO_4^- S is reduced to sulphide form by the members of bacteria belonging two genera *viz. Desulfovibrio* and *Desulfotomaculum*. These organisms utilize the combined oxygen in sulphate to oxidize the organic materials.

$$2R\text{-}CH_2OH + SO_4^{2-} \rightarrow 2R\text{-}COOH + 2H_2O + S^{2-}$$
Organic alcohol + sulphate $\rightarrow$ organic acid + water + sulphide

Organic form of sulphur: The availability of sulphur present in organic form is governed by the factors of decomposition of organic matter. In well drained soils sulphur is present as SO_4^{2-} form while in ill-drained soils in H_2S form. SO_4^- is mobile, easily lost by leaching. In arid and semiarid regions sulphur is associated with $CaSO_4 \ 2H_2O$ which accumulates in the lower horizons. SO_4^- S availability is governed by soil organic matter, SO_4^- adsorption capacity of the soil. High $H_2PO_4^-$ or HPO_4^- fertilization releases adsorbed SO_4^- S and may be lost by leaching. Sulphate adsorption is a conservation mechanism.

Elemental sulphur/sulphide sulphur is unavailable to plants. They should be oxidized to SO_4^- form sulphur for plant uptake.

$$H_2S + 2O_2 \rightarrow H_2SO_4 \rightarrow 2H^+ + SO_4^{2-}$$
$$2S + 3O_2 + 2H_2O \rightarrow 2H_2SO_4 \ 4H^+ + 2SO_4^{2-}$$

Sulphur oxidation in soil depends on the

- Microbial population in soil
- Characteristics of the sulphur source
- Soil environmental conditions

Microbial population in soil: Sulphur is oxidized by the members of the several groups of microorganisms.

Chemolithotropic bacteria which utilize the energy released during oxidation of inorganic sulphur for carbon dioxide fixation in organic matter.

They are strict autotropic bacteria *e.g., Thiobacillus thioparus, Thiobacillus coproliticus, Thiobacillus ferrooxidans, Thiobacillus thiooxidans, Thiobacillus kababis, Thiobacillus denitrificans, Thiobacillus novellus.*

Other group of bacteria is photolithotropic sulphur bacteria which carryout photosynthetic carbon fixation using sulphide and other sulphur sources. *e.g., Chlorobium* and *Chromatium.*

In addition heterotrophic organisms also oxidize sulphur to thiosulphate sulphur in the rhizosphere.

Characteristics of the sulphur source

The effectiveness of the applied sulphur source depends on the fineness of the material. Finer particles expose more surface area for microbial interaction. Hundred per cent of the sulphur material used should pass through 16 mesh and 50 per cent should pass through 100 mesh for greater efficiency. Effectiveness of the applied sulphur will be more at higher rate of application. Incorporation of the sulphur source will enhance the effectiveness as compared to localized placement as it provides better opportunity for oxidation by microorganisms and reduces local acidity.

The soil environmental requirements for sulphur oxidation are optimum soil temperature (25 °C–40 °C), soil moisture nearing field capacity, good soil aeration and acidic soil pH.

The organic sulphur transformation follows that of nitrogen. A definite ratio exists between C: N: S. The reasonable ratio being 130: 10: 1.3. Apart from soil environmental factors sulfatase activity governs the hydrolysis of esters which contain sulphur.

Critical limit of sulphur in plants

10 ppm SO_4^--S estimated using calcium dihydrogen phosphate or potassium dihydrogen phosphate extractant is considered as critical for optimum plant growth. While 13.2 ppm estimated by using calcium chloride was also considered as optimum. Besides N: S is also considered as an index of sulphur deficiency. The optimum ratio N: S being 14:1–16:1.

Based on the sulphur content of the soil, four sulphur fertility classes are delineated.

Sulphur Fertility	Available sulphur (ppm)
Very low	< 5
Low	5–10
Medium	10–15
High	15

Sulphur fertilizers

Fertilizers containing SO_4^{2-}-S

1. Single superphosphate (12–13.9% S)
2. Ammonium sulphate (24% S)
3. Ammonium phosphate sulphate (15% S)
4. Potassium sulphate (18% S)
5. Gypsum (13% S)
6. Phosphogypsum (18% S)

Pyrites (16–53.5% S) are natural minerals containing sulphur in the sulphide form. Sulphide sulphur is recommended for well aerated (non-water logged) soils 4–6 weeks before planting crop to facilitate the change from sulphide to sulphate. It is more suitable for sodic and alkali soils.

Elemental sulphur (85–100% S) is most concentrated source. It is more suitable for saline and calcareous soils. Elemental sulphur as sulphur bentonite in the form of granules or pistilles was introduced. It is a slow release sulphur source as compared to sulphate containing materials.

Table 12. Sulphur sources and sulphur content.

Sulphur sources	Sulphur content (%)
Ammonium sulphate	23.3
Ferrous sulphate	18.8
Gypsum (hydrated)	18.6
Magnesium sulphate	13
Potassium sulphate	17.6
Pyrites	53.5
Sulphur granular with additives	68–95
Single superphosphate	13.9
Urea sulphur	10–20
Zinc sulphate	17.8

QUESTIONS

Answer the following

1. What is the role of calcium in plants?
2. Which soils are deficient in calcium?
3. What are the deficiency disorders of calcium in plants?
4. What is the mechanism of calcium absorption by plants?

5. Explain calcium fertilization of crops.
6. What is the role of magnesium in plants?
7. What are the deficiency disorders of magnesium?
8. What is the relationship between calcium and magnesium, potassium and magnesium for satisfactory magnesium nutrition?
9. What are the magnesium fertilizers suitable for soil and foliar application?
10. Explain the significance of sulphur fertilization on the quality of crops.
11. Describe sulphur dynamics in soil.
12. Under what situation sulphur deficiency is observed in the soil?
13. What is the structural and functional role of sulphur in plants?
14. Describe elemental sulphur fertilization in crop production.

Give reason
1. Sulphur deficiency disorder is observed in the younger leaves of the plants
2. Gypsum application into the podding zone of groundnut is more effective.
3. Elemental sulphur should be applied two to three weeks in advance of sowing the crop.
4. Sulphur deficiency is wide spread across the country.

CHOOSE THE CORRECT ANSWER

1. The element which is constituent of chlorophyll
   (a) Magnesium   (b) Calcium
   (c) Potassium   (d) Phosphorus
2. The element involved in nitrate uptake and metabolism
   (a) Sulphur   (b) Phosphorus
   (c) Calcium   (d) Magnesium
3. Grass tetany in cattle is associated with cattle consuming forage low in
   (a) Calcium   (b) Sulphur
   (c) Molybdenum   (d) Magnesium
4. Epsom salt contains
   (a) Calcium   (b) Magnesium
   (c) Potassium   (d) Sodium
5. An element whose downward translocation in plants is limited
   (a) Magnesium   (b) Calcium
   (c) Sulphur   (d) Copper

6. The particle size of the liming material should be
   (a) 0.50 mμ or less   (b) 0.25 mμ or less
   (c) 1.00 mμ or less   (d) 2.00 mμ or less

7. Akiochi is due to toxicity of
   (a) Iron   (b) Sulphur
   (c) Manganese   (d) Boron

8. Sulphur source suitable for sodic soils
   (a) Gypsum   (b) Potassium sulphate
   (c) Pyrites   (d) Ammonium sulphate

9. The toxicity of sulphur can be alleviated by
   (a) Draining water   (b) Impounding water
   (c) Green manuring   (d) Zinc fertilizer use

Chapter 8

Micronutrients and Beneficial Elements: Role, Dynamics and Management

ZINC (Zn)

Chandler *et al.*, (1932) deduced the requirement of zinc by plants while working solution for little leaf or "rosette" on peach. Zinc plays significant role in assimilation of NO_3 and for synthesis of tryptophan. Involved in biosynthesis of indole acetic acid which is responsible for flowering and fruiting, it is an activator of several enzymes (triphosphate dehydrogenase and tryptophan synthetase), involved in photosynthesis and nitrogen metabolism.

Deficiency results in poor growth of the terminal bud. Decrease in stem length resulting in resetting and whorling of leaves. Zinc deficiency disorder in rice is called *khaira*. Common when zinc availability is < 0.7 ppm. White bud in maize and malformation of fruits, reduction in water uptake. Interveinal chlorosis often with necrosis and pigmentation and reduced leaf size in broad leaved crops known as little leaf and malformation of leaves.

The crops which are known to respond for zinc application are maize, rice, soyabean, castor, onion, barley, wheat, potato, sorghum and cotton

Zinc deficient soils
- Zinc is deficient in acid sandy soils low in total zinc,
- Calcareous soils, neutral and basic soils.
- Soils rich in silt and fine clay
- Soils high in available phosphorus
- Subsoils exposed by leveling or eroded soils

Zinc availability in soil

Zinc is present in the soil as Zn^{2+} which is water soluble and exchangeable form. Zinc is also adsorbed on the surface of clays, sesquioxides, carbonates and organic matter. Soil solution zinc is in very low concentration. Zinc forms insoluble complexes with organic matter, while zinc associated with short chain organic acids and bases is soluble. At higher pH (>7.7) zinc is present in

Zn (OH) At pH 9.1 zinc is present as Zn $(OH)_2$ which is insoluble. Zinc is not mobile in soils.

Zinc uptake is inhibited if there is excess of Cu^{2+}, Fe^{2+}, Mn^{2+} as they compete for the same cation sites. Anions like $H_2PO_4^-$ or HPO_4^{--} known to form insoluble $Zn_3(PO_4)_2$ 4 H_2O. but the solubility of this product is high to meet the crop demand. According to Lonergan in Australia and Jackson and Christensen zinc deficient plants known to lose the ability to regulate phosphorus uptake. Phosphorus accumulates in excess quantities. The phosphorus toxicity in plants resembles that of zinc deficiency. Sulphates increase zinc availability in soil. Liberal use of nitrogen increase plant demand for zinc. Zinc deficiency in plants is observed in cool and wet seasons which disappear as the temperature rises. pH induced zinc deficiency occurs within the range of 6.0–8.0, and in calcareous soils zinc is transformed into insoluble $Zn(OH)_2$ and $ZnCO_3$. Liming reduces availability of zinc.

Table 13. Crops sensitive to low levels of zinc

Very sensitive	Mildly sensitive	Insensitive
Castor	Barley	Carrot
Maize	Cotton	Crucifers
Citrus	Potato	Oats
Onion	Sorghum	Rye
Grape	Tobacco	safflower
Rice	Wheat	
Soyabean	Sugarbeet	

Zinc management

Table 14. Zinc fertilizers and zinc content.

Zinc fertilizers	Zinc concentration (%)
Zinc sulphate	23
Zinc oxide	78
Zinc carbonate	52
Zinc phosphate	51
Na_2Zn EDTA	14
Na Zn NTA	13
Na Zn HEDTA	9

The inorganic salts are applied in the range 3–20 kg per ha. Chelates are applied at 0.5–2.25 kg ha^{-1}. Clay soils need higher rates than loams and sandy soils. Most field crops need 10kg ha^{-1}. Rice crop is fertilized with 20 kg zinc sulphate per ha.

Soil application: Zinc is thoroughly incorporated into the soil as it is less mobile; band placement of zinc with nitrogenous fertilizers is useful.

Foliar application: Zinc sulphate spray at 0.5% to 1% is followed. Hydrated lime is added in equal proportion to alleviate scorching effect on zinc sulphate. (1 kg zinc sulphate and 1kg calcium hydroxide in 10 l each. Add zinc sulphate to calcium hydroxide solution and make up the volume to 100 l which is 1% zinc sulphate solution. Zinc sulphate spray at 0.5% induces resistance against brown leaf spot and bacterial blight.

Soaking seed in zinc solution: potato tubers are dipped in 2% ZnO solution.

Root dipping of seedlings: Paddy seedlings are dipped in slurry of $ZnSO_4$.

Trunk injections/driving zinc coated nails into tree trunk.

Native zinc can be mobilized by using acid forming fertilizers *viz.* ammonium sulphate, ammonium nitrate or sulphur.

IRON (Fe)

Iron plays catalytic role in the activities of several enzymes, chlorophyll synthesis, pigmentation and nitrogen fixation, photosynthesis and respiration. Its deficiency causes interveinal chlorosis in younger leaves, under acute deficient situation leaves may turn white.

Soils deficient in iron

- Calcarious soils, lime induced iron chlorosis
- Poorly drained soils, poorly aerated or oxygen deficient soils.
- Very low and very high soil temperature
- Acid sandy soils where iron chlorosis is due to copper toxicity.
- Toxicity of cobalt, chromium, manganese, nickel, copper, zinc and molybdenum. Nutrient imbalances *viz.* excess of phosphorus fertilization, potassium deficiency leads to iron deficiency.

Dynamics of iron in soils

Iron occurs in the soil as primary or secondary mineral Fe, adsorbed Fe, organic Fe and soil solution Fe. Iron in soil solution and organic iron are available to plants. Iron in solution primarily occurs as $Fe(OH)_2^+$. The mineral denoted by soil iron represents an amorphous $Fe(OH)_3$ precipitate which appears to control the solution Fe^{3+} concentration in most of the soils. For every unit increase in the pH the Fe^{3+} concentration decreases by 1000- fold. In contrast Fe^{2+} decreases by 100-fold for each unit increase in pH.

Oxidation-reduction reactions, normally the result of changes in the oxygen partial pressure, exert considerable influence on the amount of soluble iron in the soil solution. The Fe^{3+} form predominates in the well-drained soils, while levels of soluble Fe^{2+} increase significantly when soils become waterlogged. In general, lowering redox increases Fe^{2+} solubility 10-fold for each unit decrease in pe + pH, a term used to quantify the redox state in a soil.

Addition of organic matter to well drained soils can improve Fe availability. Additions of organic manure provide chelating agents that aid in maintaining the solubility of micronutrients. The presence of organic manure can increase Fe^{2+} solubility in waterlogged soils.

Iron management

Diagnosis for iron deficiency in plants: apply dilute solution of ferrous sulphate (0.5–1.0% in water) to the chlorotic leaf surface. Recovery of greenness indicates the iron deficiency in plants.

Iron compounds applied to soil readily transform into insoluble ferric form. Hence foliar application is most efficient.

Stem injection of trees with ferric carbonate or finely ground magnetite.

Iron citrate is applied to pruned cuts of grapes.

Iron chelates retain iron in available form by protecting iron from the reactions which results in formation of insoluble iron compounds *viz.* ferric hydroxide, ferric phosphate, ferric carbonate.

Iron chelates *viz.* Fe EDTA, Fe HEEDTA, Fe DTPA, Fe EDDHA are efficient in order in soils from slightly acidic to calcareous.

Table 15. Iron fertilizers and iron content

Iron sources	Iron content (%)
Ferrous sulphate	19
Ferric sulphate	23
Ferrous oxide	77
Ferric oxide	69
Ferrous ammonium sulphate	14
Ferrous ammonium phosphate	29
Iron ammonium polyphosphate	22
Iron chelates Na Fe-EDTA	5–14

MANGANESE (Mn)

Manganese is required for chlorophyll formation and development of chloroplast. Essential for respiration, nitrogen metabolism and photosynthesis, activator of variety of enzymes concerned with hydrolysis.

Deficiency: Dark green bands along the midrib and main veins with lighter green areas between the bands. In severe cases dark green areas tends to dull green and light green while lighter green area becomes white. The twig bearing such foliage shows dieback symptom. Chlorosis, necrotic spots in the interveinal areas in cereal leaves, mottling in dicot leaves. Manganese deficiency disorder is called by different names.

Gray specks–Oats

Speckled yellows –Sugar beet

Marsh spot–Peas

Pahala blight–Sugarcane

Apple, citrus, oats, sugarbeet are susceptible to manganese deficiency.

If the leaves contain < 20ppm deficient, 20–500 ppm sufficient, >1000 ppm excess.

Manganese deficient soils

• Calcareous soils, Excessive liming induces manganese deficiency.

• Very sandy acid mineral soils that are low in native manganese content,

• Soils having pH > 6.5 favour the oxidation of manganese into manganic form of manganese which is not available to plants.

• Manganese may be leached from strongly acid soils resulting in deficiency

Manganese toxicity occurs in

• Strongly acid soils

• Poorly aerated soils

• Continuous use of acid forming fertilizers

Manganese exists as solution manganese, exchangeable manganese, organically bound manganese and various manganese minerals. For satisfactory manganese nutrition of crops solution and exchangeable manganese should be 2–3 ppm and 0.2–5 ppm respectively.

The principal species of manganese in soil solution is Mn^{2+} which decreases 100 fold for each unit increase in soil pH. The concentration of Mn^{2+} in solution is predominantly controlled by MnO_2 Concentration of Mn^{2+} in the soil solution of acid and neutral soils is commonly in the range of 0.01–1 ppm with the organically complexed Mn^{2+} comprising about 90% of solution Mn^{2+}. Plants absorb Mn^{2+}, which moves to the root surface by diffusion of principally chelated Mn^{2+}. Water logging, addition of organic manures favours manganese availability in soil. Total manganese in soil is not a good indicator of availability of manganese in soils. Easily reducible manganese dioxide is of greater significance.

High levels of copper, iron or zinc can reduce manganese uptake by the plants. Manganese uptake by the crops increases with the application of $KCl > KNO_3 > K_2SO_4$.

Soil application and injections into tree trunks are followed.

Table 16. Manganese fertilizers and manganese content.

Manganese sources	Manganese (%)
Manganese sulphate	26–28
Manganese chloride	17
Manganese oxide	41–68
Chelates Mn EDTA	5–12

Alfalfa, citrus, oats, onion, potato, soyabean, sugarbeet and wheat responds to manganese application.

BORON (B)

Required for development and growth of new tissues particularly vascular elements, carbohydrate and nucleic acid metabolism, translocation of photosynthates as sugarborate complex.

Boron deficiency is associated with abnormal or retarded apical growth or resetting die- back, discoloration or failure to grow or elongate and stimulate lateral growth of plants. Young leaves become thicker, wrinkled darkish blue green colour, irregular chlorosis between midrib and interveinal region. Leaves and stems become brittle. Petioles or stems may be thickened, corky, cracked or cross hatched or may show water soaked dead areas. In fruits and tubers or roots- the fleshy parts may show brown flecks, necrosis, cracks or dry rot, may be water soaked or may show discolouration in the vascular system. Boron deficiency disorder is called by different names in different crops:

Sugarbeet–heart rot

Tobacco–top sickness

Apple–cark disease

Cauliflower–brown rot

Sunflower–malformed capitulum, with unfilled regions

Groundnut–hollow heart

Potato–brown discolouration in the vascular region

Boron is commonly deficient in sandy soils, peat and muck soils, soils low in organic matter. Alkaline soils especially those containing free lime. Excess potash application at low levels of boron intensifies boron deficiency. Boron deficiency intensifies at low levels of calcium. Moisture stress induces boron deficiency.

Boron is nonmetal anion. Plants absorb by mass flow in the form of H_3BO_3 or $H_2BO_3^{-1}$.

Boron is immobile in plants but mobile in soils.

Toxicity of boron

Tips and margins show burnt or scorched appearance. Crops sensitive to boron toxicity are grapes, kidney bean and soybean.

Boron exists in four major forms in soil: in rocks and minerals, adsorbed on clay surfaces and Fe and Al oxides, combined with organic matter and as free boric acid (H_3BO_3) and $B(OH)_4^-$ in soil solution. Boron from the soil solution to absorbing root surface is transported by mass flow or diffusion. Adsorbed form is the labile pool of boron or reserve boron which maintains boron concentration in soil solution; organically complexed boron is of significant quantity held in organic matter fraction. Soils high in organic matter are high in available boron.

The factors which influence the availability of boron are soil texture, type of clay mineral, soil pH, organic matter content, and calcium and potassium levels in the soils. High amounts of calcium in soils restrict the availability of boron; similarly high amount of potassium application when boron in soils is low induces boron deficiency.

The critical levels of boron in plants

Deficient < 1 ppm

Medium 1–5 ppm

High > 5 ppm

Boron fertility levels in soils. Hot water soluble boron is available to plants.

Deficient 15–20 ppm

Moderate 20–100 ppm

Excess > 200ppm

Boron fertilizers

Borax ($Na_2 B_4O_7 10H_2O$) (11% B)

Boric acid (17%B)

Solubor (20–21%B)

Colemanite ($Ca_2 B_6O_{11}.5H_2O$): often used in sandy soils because less soluble and less subject to leaching.

Boron is applied by

Broadcasting: 0.4–2.7 kg per ha

Foliar: 0.09–0.4 kg per ha

Dusting borax: 2–5 kg per ha

Band placement 0.4–0.9 kg per ha

Direct spray –0.2% borax solution

High boron requiring plants are crucifers, among cereals maize, oilseeds *viz.* sunflower and groundnut.

Table 17. Crop response to boron deficiency

Highly sensitive	Moderately sensitive	Low sensitive
Alfalfa	Apple	Barley
Cauliflower	Cabbage	Bean
Groundnut	Carrot	Cucumber
Sugarbeet	Cotton	Maize
Grapes	Radish	Oat
Soyabean	Tomato	Onion

COPPER (Cu)

Grossenbacher (1916) and Floyd (1917) in Florida reported that dieback of citrus could be controlled by spraying the affected trees with Bordeaux mixture but they did not interpret their results to indicate that copper was an essential element for citrus.

Copper is a part of enzymes involved in cellular oxidation- reduction, electron transport during photosynthesis and carbon assimilation and involved in several metabolic process. Soils with 31 ppm copper imparts good colour to onion bulbs while 20 ppm copper produce poor colour onion. Forage crops should be adequately manured with copper and molybdenum. Animals fed with forage deficient in copper causes various abnormalities *viz.* anemia and neural disorder. The quality of wool is reduced. So forage with low molybdenum is associated with accumulation of copper in liver (< 0.1 ppm). But high molybdenum (> 0.5 ppm) causes poisoning similar to copper deficiency. High molybdenum interferes with phosphorus assimilation and metabolism. Therefore, reciprocal relationship exists between copper and molybdenum.

Copper deficiency leads to chlorosis, dieback of terminal bud, shortened internodes, splitting of young fruits observed in orange, cabbage, lettuce; poor or absence of heading; withering and distortion of apices of younger leaves under extreme deficiency of copper in cereals and legumes. Copper deficiency occurs in histosols or organic soils. Copper is strongly held by organic acids in chelated form which are released during decomposition process of organic matter. The functional groups *viz.* carbonyl, carboxyl, phenol and amide of organic matter strongly bind copper. Soils developed on deep peats, strongly leached podzols under heather and some soils developed on limestones when reclaimed for agriculture purpose develop copper deficiency which is known as *reclamation* disease.

Copper deficient soils

- Peat and muck soils
- Alkaline and calcareous soils
- Leached sandy soils. Sandy soils have 1–30 ppm copper while loamy and clay soils posses10–200 ppm of copper.
- Leached acid soils
- Soils heavily fertilized with nitrogen and phosphorus

Organically complexed copper is in equilibrium with soil solution copper. Copper is most likely supplied to the plant roots by diffusion of organically bound, chelated copper. The copper concentration in the soil solution is very low. The dominant species of copper at pH <7 is Cu^{2+} and at pH > 7 is $Cu(OH)_2$. The $CuSO_4$ and $CuCO_3$ are also important forms of soluble copper. Solubility of copper is pH dependent. It increases 100-fold for each unit decrease in pH. Organic compounds in soil solution are capable of chelating solution Cu^{2+} which increases solution Cu^{2+}. Adsorbed copper and occluded and coprecipitated copper are not readily available for the plants.

Interaction of copper with other nutrients

Application of acid forming nitrogen fertilizers aggravates copper deficiency which is related to increase in activity of Al^{3+} levels in the soil solution. Application of nitrogen or N-P-K fertilizer results in increased growth of plants greater than copper uptake which dilutes copper in plants. Increased nitrogen application reduces the mobility of copper in plants, since high nitrogen in plants impedes translocation of copper from older leaves to new growth. High concentration of zinc, iron and phosphorus in soil may depress the copper uptake by the plants. Copper is tightly held on soil and is not subject to leaching out of the principal root zone. The amount removed by the crops is infinitesimally small compared with the amounts usually applied. Citrus requires 30 years to remove 0.89 kg per ha of copper from the soil.

Table 18. Relative response of crops to copper fertilization

Highly responsive to copper	Crops tolerant to low copper level	Extremely tolerant to very low copper level
Wheat, rice, alfalfa, carrot, lettuce, spinach, citrus and onion	Beans, peas, potato, asparagus, rye, soyabean, rape,	Rye

Soil and foliar application of copper are both effective. Soil applications are more common at copper rates 0.675–23 kg per ha. The copper fertilizer should be thoroughly mixed in the root zone. The residual effect of copper persists for 2 or more years. The foliar applications of Cu chelates are more effective to correct the copper deficiency immediately.

Foliar application of bordeaux mixture: It is prepared by dissolving 2.268 to 4.536 kg of $CuSO_4$ in 380 lit. of water followed by adding equal amount of lime or sodium carbonate.

Root dipping of seedlings in 1% Cu solution of CuO or Cu_2O is followed which is more cost effective.

Table 19. Copper fertilizers and copper content

Source of copper	Copper (%)
$Cu SO_4 H_2O$ (Monohydrate)	35
$Cu SO_4 5H_2O$ (Pentahydrate)	25
CuO (Cupric Oxide)	75
Cu_2O (Cuprous Oxide)	89
Na-Cu-EDTA	13
Na-Cu-HEDTA	9

Table 20. Organic copper fertilizers and copper content

Copper sources	Copper (ppm)
Farmyard manure	34
Pig manure	86
Poultry manure	69
Sludge of household area	130
Sludge of industrial area	1477

MOLYBDENUM (Mo)

It is the constituent of nitrate reductase, which catalyzes the conversion of NO_3^- to NO_2^- which primarily occurs in the chloroplasts of the leaves. In the absence of molybdenum nitrates are not metabolized in plants. Hence plants appear dark green. It is also structural component of nitrogenase, the enzyme actively involved in N_2 fixation by root nodule bacteria of leguminous crops, by some algae and actinomycetes, and by free- living N_2 –fixing organisms such as Azotobacter. Molybdenum also plays essential role in iron absorption and translocation.

Deficiency of molybdenum in plants causes marginal scorching, rolling or cupping of leaves chlorotic mottling between the veins of older and middle leaves all over the surface, necrotic under severe deficiency, inhibition of flower formation, nodule development and nitrogen fixation. Deficiency causes whiptail in cauliflower, yellow spot in citrus, downward cupping in radish and scald in beans. Tomato, cauliflower and rapeseed are the indicator plants of molybdenum deficiency.

Excess of molybdenum causes molybdenum toxicity in ruminants called molybdenosis. Molybdenum toxicity occurs due to excess use of molybdenum fertilizers, liming, low copper and high SO_4–S.

Molybdenum deficient soils

- Under low soil pH < 6.0 molybdenum is strongly adsorbed on the clay minerals or soil colloids.
- Molybdenum is fixed by secondary minerals like Allophone.
- Molybdenum depletion in neutral to alkaline soils by intensive cropping.
- Deficient in podzolized soils.

Molybdenum in soil include nonexchangeable Mo in primary and secondary minerals, exchangeable Mo held by iron and aluminium oxide, Mo in soil solution and organically bound Mo. Mo in soil solution occurs predominantly as MoO_4^{2-}, $HMoO_4^{-}$ and H_2MoO_4 The concentration of MoO_4^{-} and $HMoO_4^{-}$ increases with increase in pH. The solubility of Mo is controlled by soil Mo which is very close to the solubility of $PbMoO_4$ or wulfenite. Plants absorb Mo as MoO_4^{2-}. Molybdenum is transported to plant roots by mass flow, while diffusion to plant roots occur at levels <4 ppb.

Factors affecting Mo availability

Mo availability increases with increase in pH of the soil.

Strongly adsorbed to Fe and Al oxides, a portion of which becomes nonavailable to plants

Phosphorus enhance MoO_4^{2-} absorption

SO_4^{2-} in soil solution depresses Mo absorption by the plants

Cu and Mn reduce Mo uptake

Mg will increase Mo uptake

NO_3–N increase Mo uptake and availability in soil.

The reasonable level of molybdenum in soil is 2.5 ppm

Molybdenum management

- Seeds are treated with 2–4g of ammonium molybdate per kg of seed. Slurry or dust of ammonium molybdate is more effective.
- Spraying 0.1% solution of sodium molybdate or ammonium molybdate is followed to correct the deficiency of Mo.
- Small quantities of molybdenum sources are sometimes combined with NPK fertilizers.
- Soaking the seeds in the solution of sodium molybdate before sowing seeds
- Liming acid soils.

Table 21. Molybdenum fertilizers and molybdenum content

Molybdenum fertilizers	Molybdenum content (%)
Sodium molybdate	38
Ammonium molybdate	54
Molybdenum trioxide	66 or less
Phosphate rock	17 ppm
Sewage sludge	2–10 ppm

Beneficial elements

Beneficial elements are those which are beneficial to some crop plants at very low concentrations. Their essentiality is not unequivocally established.

- They are able to positively affect the uptake, translocation and utilization of essential elements
- They activate certain enzymes affecting the production of an essential metabolite
- They may counteract the toxic effect of some other element or antimetabolite
- They may impact disease tolerance, stress resistance
- They may improve the quality and yield of crops.
- They may provide strength to stem to withstand lodging in cereals.

Sodium (Na): beneficial for growth of some plants (halophytic plants) spinach, sugarbeet, turnip. Favourable effects of sodium are also reported to occur in cabbage, mustard, radish and rape seed. The increased growth produced by salt in halophytes is believed to be due to increased turgor. Many plants that possess C_4 dicarboxylic pathway require sodium as an essential nutrient. It also has role in inducing crassulacean acid metabolism. Water economy is related to C_4 plants.

Table 22. Sodium uptake potential of crops

High	Medium	Low	Very low
Fodder beet	Cabbage	Barley	Buck wheat
Sugarbeet	Coconut	Flax	Maize
spinach	Cotton	Millet	Soyabean
	Oats	Rapeseed	Rye
	Potato	wheat	
	Rubber		
	turnip		

Table 23. Crop response to sodium when potassium is deficient or absent

When potassium is absent responds to sodium	When potassium is deficient responds to sodium	No response to sodium, requires potassium
Cabbage	Barley	Corn
Radish	Oats	Potato
	Wheat	
	Cotton	

In some crops sodium demand appears to be independent and perhaps greater than potassium demand. Sodium fertilization of forage crops is considered desirable. Sodium concentration between 1–2% in pasture grasses will improve palatability and will provide part of the animal sodium requirement. Sodium fertilizer sources are sodium chloride and sodium nitrate.

Silicon (Si)

Silica involved in root functions contributes to the drought tolerance of crops *e.g.*, in sorghum.

Silica impregnates into the wall of epidermal and vascular tissues, where it appears to strengthen the tissues, reduces water loss, retard fungal infection, resistance to lodging. Regulate photosynthesis and enzyme activity in sugarcane. It increases the activity of invertase in sugarcane resulting in greater sucrose production. Reduction in the phosphatase activity in sugarcane is believed to provide greater supply of essential high energy precursor needed for optimum cane growth and sugar production. Corrects the toxicity of manganese, iron, active aluminium, prevents localized accumulation of manganese in sugarcane leaves.

Freckling, a necrotic leaf spot condition is a symptom of low silica in sugarcane receiving direct sunlight. Ultraviolet radiation seems to be the causative agent in sunlight since plants kept under plexiglass or glass does not freckle. Which suggests the silica in sugarcane filters the harmful ultraviolet radiation.

Silica in rice maintains erectness of leaves there by increases solar radiation interception and photosynthesis. The oxidizing power of rice roots and accompanying tolerance to iron and manganese toxicity are related to silica nutrition. Reduces transpiration, reduces water stress inside the plant. Reduced uptake of sodium, chlorine and magnesium, reduced lodging due to excess nitrogen application. Reduced incidence of blast, brown spot, leaf scald, sheath blight and grain discolouration.

Silicon deficient soils exist in intensively weathered, high rainfall regions. Properties of the silicon deficient soil include low total silica, high aluminium, low base saturation, low pH, high phosphorus fixing capacity due to their high anion exchange capacity and aluminium and iron content. Plant available Fe^{2+} and Al^{2+} is more in these soils. Silicic acid is the principal silicon species in solution. H_2SiO_4 polymerizes to form precipitates of amorphous silica (SiO_2). Silica is absorbed by the plants as monosilicic acid ($Si(OH)_4$). Cereals and grasses contain 0.2–2.0 per cent silica. Sodium acetate (NaOAc) extractable Si rated as adequate for rice production is >130 ppm in Japan and Korea, and > 90 ppm in Taiwan. Concentration of silicon in soil solution is largely controlled by a pH dependent adsorption reaction. Silicon is adsorbed on the surfaces of Fe and Al oxides. Solution concentration of silicon increases with time of submergence which encourages silica uptake by rice. Silicon bearing materials are added when high rates of nitrogen fertilizers are used. Low land rice contains 4.6–7.0 per cent Si in straw. Rice yielding 5 tonnes per ha accumulate 230–470 kg of Si ha^{-1}. Sugar yields from sugarcane increased in Hawaii with the application of electric furnace slag. Annual application of 566–1132 kg per ha calcium silicate in rows improved the sugarcane yield. Silica fertilization is known to increase the rice yield by 10–30 per cent. The silica may be supplied through electric furnace calcium silicate slag (18–21% Si), Lignite fly ash (23% silica). Rates of 1.5–2.0 t ha^{-1} of silicate slag provide sufficient silica for rice produced in low silicon soils.

Table 24. Silicon fertilizers and silicon content

Silicon fertilizers	Silicon (%)
Calcium silicate slag	18–21
Calcium silicate	31
Sodium metasilicate	23

Vanadium (Va)

Vanadium may substitute for molybdenum in nitrogen fixation by microorganisms such as the rhizobia. Increase in growth attributable to vanadium has been reported in asparagus, rice, lettuce, barley and corn.

Vanadium requirement of plant is said to be < 2 ppb on dry weight, where as normal concentration in plants averages about 1 ppm.

Cobalt

Cobalt is essential for symbiotic nitrogen fixation in some microorganisms and in synthesis of vitamin B_{12} in ruminant animals. Cobalt deficiency in ruminant animals is associated with forages produced from soils containing < 5 ppm of total cobalt. Cobalt (Co) is adsorbed on the exchange complex as clay-organic matter complexes. The crystalline manganese oxide minerals in the soil have high adsorption capacity of cobalt. Presence of these minerals reduce the cobalt availability in soil. Normal concentration of cobalt in plants ranges from 0.02–0.5 ppm. Response to cobalt has been reported in cotton, beans and mustard. 10 ppb of cobalt in nutrient solution was sufficient for nitrogen fixation by alfalfa. The forage crops grown in soils deficient in cobalt are fertilized with cobalt at 106.25–212.5 g per ha as cobalt sulphate. Satisfactory nodulation in clover and alfalfa was achieved in Australia by fertilizing 35.42–141.75 g cobalt sulphate per ha. Small amount of cobalt sulphate may be mixed with superphosphate and used to increase cobalt concentration in subterranean clover.

Nickel

Plant content of nickel ranges from 0.1–1.0 ppm on dry weight. Nickel is a component of urease enzyme catalyzes hydrolysis of urea. Legumes are known to respond for nickel. In soyabean nodule weight and seed yield increase have been reported with Nickel(Ni) fertilization. Application of some sewage sludges may result in elevated levels of nickel in plants.

Selenium

Selenium requirement is important in human and animals. Deficiency in human causes cardiomyopathy. Problems of chronic arthritis, deformity of joints, osteoarthropathy are common.

The available selenium concentration in soil is 0.3 ppm. Selenite (SeO_3^{2-}) in acid soils, Selenates (SeO_4^{2-}) in well aerated soils of arid and semiarid regions. Organic Selenium(Se) is important, since upto 40% total selenium is present in humus. Soluble selenium compounds are liberated through the decay of seleniferous plants. Organic selenium is stable and available to the plants. Selenium is deficient in most of the cereals except rice.

Application of sodium selenite at 70.875 g per ha is satisfactory for forages. Foliar application of sodium selenite at 15 g Se per ha is efficient to increase selenium content in maize. Selenium is present in rock phosphate and superphosphate produced from them containing 20 ppm or more selenium may provide sufficient selenium required by the plants.

Selenium at 6g per tonne of fertilizer is blended in Finland and Newzealand to meet the selenium requirement of crops.

QUESTIONS

Answer the following

1. What is the role of zinc in plants?
2. Which soils are deficient in zinc?
3. Explain zinc and phosphorus interaction in soil and plants.
4. Describe zinc fertilizer management in crops.
5. What are the deficiency disorders of zinc in plants?
6. What is the role of iron in plants?
7. What are the deficiency disorders of iron?
8. Which soils are deficient in iron?
9. Describe iron management in crops.
10. What is the dynamics of iron in soils?
11. What is the role of manganese in plants?
12. What are the deficiency disorders of manganese?
13. Which are the manganese deficient soils?
14. Why soil application of manganese is not effective on crops?
15. What are the deficiency disorders of boron in plants?
16. What are the situations under which boron is deficient in soils?
17. State the role of boron in plants.
18. Explain the transformations of boron in soils.
19. Explain the boron fertilization in crops.
20. What are the factors which influence the availability of boron in soils?
21. What are the functional roles of copper?
22. What are the deficiency symptoms of copper in plants?
23. What is the interrelation between copper and soil organic matter?
24. Which are the copper deficient soils?
25. What are the forms of soil copper?
26. What is the interaction of nitrogen fertilization and copper in plants?
27. What are the efficient methods of copper fertilization in crops?
28. Explain molybdenum fertilization of crops.
29. What are the deficiency disorders of molybdenum?

30. Which soils are deficient in molybdenum?
31. What are the factors of molybdenum availability in soils?
32. What are the roles of beneficial elements in crop plants?
33. What are the beneficial roles of silicon in sugarcane?
34. What are the beneficial roles of silicon in rice?
35. Which soils are deficit in silicon?
36. Explain silicon fertilizer management in crops.

Short answers

1. What form of silicon is absorbed by the plants?
2. What is the satisfactory level of silicon in soils for satisfactory crop production?
3. What is the quantity of silica removal by rice and sugarcane?
4. What are the cobalt fertilizers?
5. What is the significance of cobalt fertilization of forages?
6. What is the role of nickel in plants?
7. What is the significance of selenium in human health?
8. Which cereal is not deficient in selenium?
9. Why cobalt and selenium are blended with fertilizer?
10. How the selenium content of the plants is increased?

Give reasons

1. Iron is deficient in calcareous soils.
2. Soil application of iron fertilizers is not effective.
3. Reclamation disease
4. Why lime is used in preparation of bordeaux mixture?

CHOOSE THE CORRECT ANSWER

1. *Khaira* is the deficiency disorder of
   (a) Copper   (b) Zinc
   (c) Molybdenum   (d) Manganese
2. Zinc sulphate contains
   (a) 23% Zn   (b) 46% Zn
   (c) 33% Zn   (d) 13% Zn
3. An element involved in biosynthesis of IAA
   (a) Manganese   (b) Iron
   (c) Zinc   (d) Boron

4. Ferrous sulphate contains
   (a) 14% Fe   (b) 19% Fe
   (c) 23% Fe   (d) 30% Fe

5. Intervenal chlorosis in younger leaves is due to
   (a) Calcium   (b) Iron
   (c) Magnesium   (d) Zinc

6. Pahala blight of sugarcane is the deficiency disorder of
   (a) Magnesium   (b) Manganese
   (c) Iron   (d) Zinc

7. Manganese sulphate contains
   (a) 30–32% Mn   (b) 26–28% Mn
   (c) 18–22% Mn   (d) 15–18% Mn

8. Increase in pH >6.5 decreases the availability of
   (a) Magnesium   (b) Calcium
   (c) Molybdenum   (d) Manganese

9. Hallow heart of groundnut is due to deficiency of
   (a) Boron   (b) Iron
   (c) Calcium   (d) Magnesium

10. Brown discoloration in the vascular region of potato is due to deficiency of
    (a) Iron   (b) Boron
    (c) Calcium   (d) Magnesium

11. Critical level of boron in plants is
    (a) < 5 ppm   (b) < 1 ppm
    (c) < 10 ppm   (d) < 15 ppm

12. Soils are said to be deficient in boron if hot water soluble boron tests
    (a) < 10 ppm   (b) < 20 ppm   (c) < 5 ppm   (d) < 50 ppm

13. Boron content of borax is
    (a) 6%   (b) 11%   (c) 15%   (d) 20%

14. An element involved in translocation of photosynthates
    (a) Manganese   (b) Iron
    (c) Zinc   (d) Boron

15. Crop highly sensitive to boron is
    (a) Maize   (b) Sunflower
    (c) Potato   (d) Grapes

16. The concentration of borax spray solution is
    (a) 2%   (b) 0.2%
    (c) 5%   (d) 0.02%

17. Crop which is highly responsive to copper fertilization
    (a) Citrus   (b) Potato
    (c) Soyabean   (d) Rye
18. Good colour is imparted to onion bulb by adequate supply of
    (a) Molybdenum   (b) Copper
    (c) Iron   (d) Zinc
19. Anemia and neural disorders in cattle are associated with forage deficient in
    (a) Zinc   (b) Molybdenum
    (c) Copper   (d) Iron
20. Copper content of copper sulphate ($CuSO_4$ $5H_2O$) is
    (a) 75% Cu   (b) 35% Cu
    (c) 25% Cu   (d) 89% Cu
21. The organic manure rich in copper is
    (a) Pig manure   (b) Cattle manure
    (c) Sheep manure   (d) Poultry manure
22. Dark green colour with no progress in growth of plants is due to deficiency of
    (a) Copper   (b) Molybdenum
    (c) Zinc   (d) Iron
23. Micronutrient which is a component of nitrogenase enzyme
    (a) Molybdenum   (b) Zinc
    (c) Boron   (d) Manganese
24. Whiptail of cauliflower is due to deficiency of
    (a) Iron   (b) Zinc
    (c) Molybdenum   (d) Copper
25. Toxicity of molybdenum causes disorders in ruminants similar to deficiency of
    (a) Zinc   (b) Copper
    (c) Iron   (d) Manganese
26. The satisfactory level of molybdenum for crop growth is
    (a) 2.5 ppm   (b) 1.0 ppm
    (c) 5.0 ppm   (d) 10.0 ppm
27. Molybdenum content of sodium molybdate is
    (a) 54%   (b) 38%
    (c) 48%   (d) 28%

28. Palatability of forage crops is improved by fertilization with
    (a) Calcium   (b) Sodium
    (c) Magnesium   (d) Nitrogen
29. The elements which impart turgor in plants are
    (a) Phosphorus and potassium   (b) Sodium and potassium
    (c) Nitrogen and phosphorus   (d) Nitrogen and potassium
30. Vanadium substitutes the function of
    (a) Copper   (b) Molybdenum
    (c) Zinc   (d) Cobalt
31. Nickel is the component of
    (a) Nitrogenase   (b) Phosphatase
    (c) Urease   (d) Nitrate reductase
32. Plant content of nickel is
    (a) 1–2 ppm   (b) 2–3 ppm
    (c) < 1 ppm   (d) 3–4 ppm

Chapter 9

Organic Manures: Production and Enrichment

Farmyard Manure is the manure prepared by using the mixture of dung and urine soaked litter by microorganisms.

Good farmyard manure is one
- In which the raw materials from which the manure is prepared will not be identifiable.
- Brown in colour. Black colour indicates anaerobic decomposition.
- Friable: should pass between the fingers
- Free from foul smell
- Contains 0.5% N, 0.25 P_2O_5 and 0.5 to 1.0% K_2O.

Compost is the synthetic farmyard manure prepared by using the decomposable plant wastes.

Composting is a way of storing the available organic materials. One cannot use the different organic matter as and when they available.

Materials with wider C: N may not decompose. Further there may be nitrate depression.

Undecomposed organic matter will not be homogenous and will have very little impact on the soil properties.

Advantages of composting: by composting the organic wastes we can
- Control foul smell
- Kill the pathogens and control the flies
- Kill weed seeds
- Minimize nitrogen loss

Loss of plant nutrients during collection and storage of on farm available nutrients sources:
There is a potential of producing 3 m t of compost annually in the country which will provide about 20.5 kg N, 11.0 kg P_2O_5 and 25 kg K_2O per tonne of compost on dry weight basis. Water hyacinth an aquatic weed is a good

substrate for compost production. It is high in potassium more suitable for high potassium requiring crops *viz.* rice, potato, maize, jute and vegetables.

NADEP method of composting

Narayana Rao D. Pandaripande popularly known as 'Nadep kaka' a believer in Gandhian philosophy has advocated this method of composting through the minimum use of cattle dung.

Materials required
- Farm wastes/crop residues: 1400–1500 kg
- Cattle dung: 90–100 kg
- Soil : 1750 kg
- Water: 1500–1700 litre

This is an aerobic method of composting. The time required for composting is 90–100 days. The compost contains 0.5 to 1.5% N 0.5–0.9% P_2O_5 and 1.2–1.4% K_2O.

Methodology

Select the site near a cattle shed or near to the farm. Tanks of size 10 ft × 6 ft × 3 ft are constructed with 9 inch thick walls. Lining of the bricks is made with the help of mud which saves expenditure on cement. For circulation of air 7 inch × 4 inch holes are left on all the four sides of the wall.

Method of filling: slurry of cattle dung is sprinkled on the floor and wall of the tank.

Crop residues/farm wastes are spread in a layer on the floor of the tank to a height of 6 inches. This forms the first layer.

Cattle dung at the rate of 4–5 kg mixed in 125–150 l of water to make slurry. This was sprinkled over the first layer. This may be treated as the second layer. Over this layer dried sieved soil about 50–60 kg is spread on the moist layer of farm refuses. Then water is sprinkled to moisten the material (third layer). Thus the tank is filled repeating the layers in sequence till the material is 1½ ft above the surface of the tank. A temporary roof using thatched material may be provided to prevent heating by sun.

After 15–20 days of filling the material shrinks due to the process of softening and compaction. About ¼ volume above the surface of the last layer is available. Again the empty space is filled with the materials in the order explained earlier. The surface is sealed and plastered with mud and dung. It is sprinkled with water and dung to maintain 15–20% moisture. The compost will be ready in 90–100 days. The composted material may be sieved using 35 mesh/ft sieve.

Manure yield: 3.0–3.5 tonnes

Cost economics
Estimated cost of the tank ₹6,000/-

Labour requirement: 8 man days

Turning the material during composting is not required.

Materials required for construction of composting structure
Bricks required (Nos.) 1500. Cement one bag is sufficient if the top layers alone are plastered. If all the walls are plastered then 4 bags are required.

Site selection
- Elevated place or avoid areas prone to water stagnation
- Under the shade of the trees.
- Near the field
- Length of the tank should be across the direction of wind

Japanese or VAT method

Tanks of 7–10 m length × 1.0–1.2 m breadth and 1 m height are constructed using stone slabs of 4ft length × 2 ft breadth and thickness 2 inches. A gap of 5 cm is left between the slabs. The floor of the tank is made impervious by using brick bats, sand and cement concrete.

Substrates

Organic substrates available with the farmer.

Livestock and poultry excreta

Organic wastes *viz.,* leaf, litter, coconut fronds, stubbles, stalks, legume residues *etc.*

By products of agrobased industries *viz* press mud, vegetable waste

Rock phosphate or ash

Microbial culture viz. *Aspergillus sp., Penicillium sp., Trichoderma viridae, Agrobacterium radiobacter, Azotobacter, Bacillus subtilis, pseudomonas sp., Spiralis, Pleurotus sp. and Polyporus sp.* In the absence of cultures well-decomposed compost or dung may be used.

Filling the tank

The tank is filled layer by layer in the following sequence leaving one meter space at one end of the tank to facilitate turning.

Sequence of layers

1ˢᵗ layer   Lignin and cellulolytic materials to a height of 10–15 cm.

2nd layer   Dry grass, stubbles, cellulolytic materials to a height of 10–15 cm. Sprinkle decomposing microbial culture, dung and urine or biogas slurry to moisten both the layers. Ash and soil are also sprinkled.

3rd layer   Nitrogen rich materials *viz.*, legume residues, weeds to a height of 10–15 cm.

4th layer   Phosphorus and potassium rich materials certain weed species *e.g. Calatropis*, rock phosphate, ash, stubbles, crop residues. Sprinkle slurry of dung/urine/biogas slurry @ 5–10 kg in 40–50 lit water.

5th layer   Residues of sunflower, redgram or maize are cut into 10–15 cm, mixed with green biomass and filled to a height of 10–15 cm.

6th layer   Dung spread to a thickness of 20–30 cm. well decomposed compost, ash or tank silt can be used.

Microbial cultures may be mixed in water and sprinkled over the paddy husk, which is uniformly sprinkled over celluloytic, lignin materials, viz., crop residues. If cultures are not available aged compost and jaggery solution (4-5%) may be used.

These six layers form a set, which are repeated till the tank is filled to a height of 40–50 cm above the surface of the tank in a doom shape. Turnings are given starting from 15th day of filling at an interval of 15–20 days. Rock phosphate and gypsum at 50–100 kg per tonne of compost is added at the time of first turning. At the last turning finished compost may be enriched with oil cakes *viz.* neem/castor/pongamia.

The compost will be ready in 90–100 days.

Preparation of vermicompost

Tanks of dimension 7–10 m length, 1.5–2.0 m width and 0.3–0.5 m height are constructed using bricks or stone slabs. Old asbestos sheets or wooden planks can also be used. The floor of the tank is made impervious and grade is provided in one direction with a drainage hole of 5 cm diameter. Tanks should be constructed under shade or shading may be done by thatched roof of coconut fronds or bamboo mat.

Organic materials are filled into the tanks till the surface level. Size reduction of large materials is desirable. Over this dung or gobar gas slurry is sprinkled. Holes of 5–10 cm diameter are made from the surface for aeration. After 2–3 weeks young worms are let into the substrate through the holes. One kg of adult worms feed 5–10 kg of substrates. Depending on the quantity of substrates the required quantity/number of worms may be introduced. Sprinkle water regularly to keep the material under moist condition. The materials are gently turned at intervals for aeration. Earthworms feed on the substrates and leave vermicasts. The compost will be ready in 50–60 days. The contents are removed from the tank and heaped on a hard surface. As the heap dries up the worms move down to the moist layer at the bottom. The dried compost is sieved to separate unutilized substrates, immature worms and egg cases. The contents retained on the sieve are mixed with the next batch of substrates during filling.

Table 25. Nutrient content of composts prepared from different methods

Type of compost	N	P_2O_5	K_2O
NADEP compost	0.5–1.5	0.5–0.9	1.2–1.4
Japanese compost	0.5–2.0	1.0–2.5	1.0–2.0
Vermi compost	0.5–1.5	0.1–0.3	0.2–0.6

Concentrated organic manures

Oil cakes: Oil cakes are the by products of the oil industry. Non-edible oil cakes can be used as manure. The nitrogen content varies with the type of oil cakes. The oil cakes are slow acting as the oil present in it prevents quick conversion of organic form of nutrients to inorganic form.

Table 26. Nutrient content (%) of oils cakes

Oil cakes	N	P_2O_5	K_2O
Castor	4.3	1.8	1.3
Neem	5.2	1.0	1.4
Cotton seed	3.9	1.8	1.6
Pongamia	4.2	0.9	1.2
Ippi (Mahua)	2.5	0.8	1.8
Cotton seed (decorticated)	6.4	2.9	2.2

Blood meal: Blood from the animals slaughtered for meat purpose is collected. It is treated with cupric sulphate at the rate of 125 g per 100 kg blood and allowed to dry. The dried blood is called blood meal. It contains 10–12% of nitrogen and 1–2% of phosphoric acid.

Meat meal: Meat from the dead animals unfit for consumption is dried and converted into meat meal.

Fish manure: The excess of fish catch during glut in market are dried and sold as fish manure. It contains 4-10% N, 3–9% P_2O_5, 0.3–1.5% K_2O, but the composition is subjected to considerable variation depending upon the kind of fish and quantity of sand mixed during drying. Fish manure contains variable quantities of oil which is readily used by the soil bacteria and the manure ingredients become readily available for the use of crops.

Raw bone meal: The bones of the slaughtered animals are ground in the natural state known as raw bone meal. It contains 3–4% N and 20–25% P_2O_5.

Steamed bone meal: bones are steamed under pressure for extraction of the constituent fat which is used in soap and candle manufacture. The nitrogen in bones is also extracted under pressure for the production of glue. The residue left is ground which is called steamed bone meal. It contains 1–3% N and 22–24% P_2O_5.

Horn and hoof meal: The horns and hoofs of the slaughtered animals is collected, powdered and used as manure. It contains 13% N and 1.5–3.0% P_2O_5.

Enrichment of bulky organic manures

Advantages of enrichment of bulky organic manures are:

- Increase the nutrient content of organic manures.
- Reduces the bulk to be handled to supply unit quantity of nutrient.
- Make available the insoluble nutrient sources.
- Prevents temporary unavailability of nutrients due to immobilization.
- Improves the nutrient use efficiency.

Enrichment during decomposition

When fertilizers are added during decomposition of organic wastes the added nutrient elements (nitrogen and phosphorus) are immobilized in the microbial body and inserted into the molecules of humic substances formed during decomposition. Inorganic nutrient elements may be adsorbed on the organic exchange sites or chelated by humic substances. Thus the added fertilizer will be transformed into a slow release fertilizer.

Nitrogen enrichment: Materials with wide C: N ratio like straw, trash or coir pith *etc.* need additional 0.7–0.8 kg mineral N/100kg raw material. Concentrated nitrogen sources may be used for enrichment. By enrichment the C: N will be narrowed. The finished product will have 1.8–2.5% N. when C: N of the raw material is narrow the enriched nitrogen will be prone to loss by volatilization. The nitrogen content of compost cannot be increased beyond 2.0–2.5% with the addition of mineral nitrogen during composting. If the compost or farmyard manure has C: N ratio 20:1 it may be brought down by adding concentrated nitrogen sources. The nitrogen content may be increased to 5–7%.

Phosphorus enrichment: Concentrated phosphorus sources may be used for enrichment. It is more rationale to use the insoluble phosphorus sources as enriching material. Phosphorus is not readily lost during enrichment process. Rock phosphate, raw bone meal and pyrites are used for enrichment.

Partially decomposed farmyard manure may be enriched with phosphorus to supply the required quantity of phosphorus to the crop.

Phosphocompost is the phoshphorus enriched organic manure. To prepare 1 tonne of phosphocompost on dry weight basis the raw materials required are given in Table 27.

Table 27. Substrates required for phosphocompost preparation.

Raw materials	Quantity (t)
organic wastes *viz.* crop residues, weeds, left over fodder *etc.*	0.8
cattle dung or biogass slurry	0.1
soil	0.1
well decomposed farmyard manure or compost or sewage sludge	0.05
Mussoorie rock phosphate	0.256

Table 28. Nutrient composition (%) of phosphocompost

Components	Compost	Phosphocompost
Organic carbon	24.3	18.6
Total nitrogen	1.15	0.82
Total phosphorus	0.21	3.48

The phosphorus in the phosphocompost is in the form of citrate soluble phosphorus.

The finished compost may be enriched with phosphorus to supply the required quantity of phosphorus to the crops.

Zinc enrichment: Transformation of available zinc into insoluble complexes may be minimized by zinc enrichment. It forms chelates with organic acids during decomposition. It is convenient for application.

Enrichment with microbial culture: Enrichment of compostable material with *Azotobacter chroococcum* increased the nitrogen content of compost by 10 to 25%. *Aspergillus awamorii* or *Bacillus polymixa* inoculation will enhance solubilization of insoluble phosphorus. Microbial cultures are enriched when the temperature of the composting material stabilizes around 30–35°C.

Enrichment method

The organic amendments, mineral additives and microbial inoculents can be enriched during composting for improving the quality and quantity of the compost produced.

1. **Blending with Organics:** During composting of the city garbage or solid wastes, certain organic substances that are rich in nitrogen and other nutrient elements can be mixed. The organic substances that can

be added are: oil extract cakes, food processing wastes, press mud from sugar industry, legume plants and green weeds, leafy materials, poultry manure, silk worm wastes/litter.

2. **Mineral additives:** Different minerals like rockphosphate can be added layer by layer during composting to get phosphorus enriched compost. The level of rockphosphate may be 12.5%–25% depending on the phosphorus enrichment. Along with addition of rockphosphate, phosphate solubilizing microorganisms either *Bacillus megaterium* or *Aspergillus awamorii* is inoculated at the rate of one or two kg per ton of wastes.

3. Microbial inoculation during composting is done for two reasons.

i. For accelerating the decomposition process by inoculation of lignin and cellulolytic fungi e.g., *Pleurotus sajur caju*. These cultures are added for each layer of the garbage or organic wastes at 2 kg/ton.

ii. Microbial inoculation for quality improvement. The nutrient status of the compost is improved by inoculation of bacterial and fungal cultures. Bacterial inoculants such as *Azotobacter, Azospirillum* are added to the decomposing materials after a period of 30–45 days of initial composting. These nitrogen-fixing cultures increase in population and enhance the nitrogen content. The fungal (*Trichoderma sp.*) or Bacterial culture (*Bacillus spp.*) are inoculated at the rate of 2 kg/ton after 30–45 days of initial degradation. These cultures multiply and improve the quality of the final product.

QUESTIONS

Define

(a) Compost

(b) Farmyard manure

Answer the following

1. What are the substrates required for NADEP composting?
2. Illustrate the method of filling NADEP compost tank.
3. What are the substrates required for VAT composting?
4. Illustrate the method of filling of VAT compost.
5. State the advantages of NADEP compost.
6. Describe the preparation of vermicompost.
7. What are the concentrated organic manures?
8. What are the advantages of enrichment of bulky organic manures?
9. What is phosphocompost?

10. Explain microbial enrichment of organic manures.

11. Describe nitrogen enrichment of bulky organic manures.

CHOOSE THE CORRECT ANSWER

1. Lignin and cellulolytic materials are degraded by
   (a) *Trichoderma*   (b) *Pleurotus sajur caju*
   (c) *Azatobacter*   (d) *Aspergillus awamorii*

2. The standard size of NADEP compost tank is
   (a) 12ft × 6ft ×3ft   (b) 10ft × 6ft ×3ft
   (c) 15ft × 4ft × 3ft   (d) 20ft × 10ft × 5ft

3. The maximum nitrogen content of fish manure is
   (a) 10%   (b) 5%   (c) 2%   (d) 1%

4. Efficient phosphorus solubilizing microorganism is
   (a) *Azotobacter* sp.   (b) *Azospirillum* sp.
   (c) *Aspergillus awamorii*   (d) *Pleurotus sajur caju*

5. The nitrogen content of oilcakes is
   (a) 4–5% N   (b) 0.5–1.0% N
   (c) 1–2% N   (d) 7–8% N

6. The $N:P_2O_5:K_2O$ supplied by one tonne of compost is
   (a) 40-20-50 kg per ha   (b) 10-5-12 kg per ha
   (c) 30-16-37 kg per ha   (d) 20-11-37 kg per ha

7. The $N-P_2O_5-K_2O$ content of farmyard manure is
   (a)0.2-0.1-0.5 per cent   (b) 0.5-0.25-1.0 per cent
   (c) 1.0-0.5-2.0 per cent   (d) 2.2-1.1-4.5 per cent

8. Labour and cattle dung requirements for manure preparation is less in
   (a) NADEP   (b) VAT
   (c) Vermicompost   (d) Farmyard manure

9. The substrate used for increasing the nitrogen content in VAT composting
   (a) Cereal residue   (b) Legume residue
   (c) Cattle dung   (d) Ash

10. The C:N ratio of the finished compost is
    (a) 20:1   (b) 30:1
    (c) 10:1   (d) 5:1

Chapter 10

Green Manures and Crop Residues: Production and Management

Green manure is defined as fresh organic matter/undecomposed plant material/ real organic matter, natural plant food added to the soil for the purpose of supplying plant nutrients.

CLASSIFICATION OF GREEN MANURES

Nonlegumes: Rape seed, buck wheat, niger, rye, oats and maize

Legumes:

Grain legumes: cowpea, mungbean, pigeon pea, soybean, khesari (*Lathyrus sativus*)

Dual purpose legumes: cluster bean (*Cyamopsis tetragonoloba*). pillipesara, *Sesbania speciosa*.

Non grain legumes: sunnhemp, dhaincha, pillipesera (*Phaseolus trilobus*), stylosanthes and desmodium.

Woody legumes: glyricidia, pongamia, leucaena.

Non leguminous woody plants: Neem (*Azadirecta indica*).

Stem nodulating: *Sesbania rostrata, Aeschynomene afraspera* can tolerate water logged condition. Nodules of *Sesbania rostrata* are arranged in four rows on the branches while that of aeschenomene are randomly distributed. Stem nodulating rhizobium are fast growing and possibly grouped under new nitrogen fixing *Azorhizobium*. The nodules *Aeschynomene* are slow growing grouped under Rhizobium. They are photo and thermosensitive.

Green manures for drought situations: Pillipesara, sunnhemp, blackgram greengram, gaur, cowpea, *Sesbania canabina* (Dhaincha), *Sesbania speiciosa*, *Sesbania macrocarpa, Sesbania sericea, Phaseolus semierectus, Phaseolus trilobus, Calapogonium. usaramoensis, Calapogonium mucunoides, Cassia leschenaultiana*.

Green manure for waterlogged situation: *Aeschynomene americana, Phaseolus semierectus, Sesbania cannabina, Sesbaia speciosa, Sesbania macrocarpa, Sesbania sericea.*

Two types: Based on the site of greenmanure production they are classified as *in situ* or *in situ* composting and *ex situ* green manures.

The crop plants selected for *in situ* green manures should be

• Multipurpose
• Short duration
• Fast growing
• High nutrient accumulating
• Tolerant to flood, shade, drought, pest, disease and weed suppressing
• Wide adoptability, high water use efficiency
• Early onset of biological nitrogen fixation
• Photoperiod insensitive
• Soft and easy for incorporation
• High seed multiplication and longer seed viability
• Response to microbial inoculation
• Wide ecological adoptability

Advantages of Green Manuring
• Adds organic matter
• Conservation of nutrients.
• Mobilization of nutrients from the deeper layers.
• Phosphorus availability increases
• Leguminous green manures improve the availability of nitrogen from humus known as priming action.
• Weed smothering effect.
• Controls root knot nematode
• They serve as catch, forage, shade, cover or fibre crops

Table 29. Potential of Green manure crops

Crop	Season	Biomass Yield (t/ha)	N yield Kg per ha	N content (%) on fresh weight basis
Sunnhemp	Kharif	21.2	91	0.43
Dhaincha	Kharif	20.2	86	0.43
Pillipesara	Kharif	18.3	201	1.10
Cowpea	Kharif	15.0	74	0.49

Crop	Season	Biomass Yield (t/ha)	N yield Kg per ha	N content (%) on fresh weight basis
Guar	Kharif	20.0	68	0.34
Senji (*Melilotus alba*)	Rabi	28.6	163	0.57
Khesari	Rabi	12.3	66	0.54
Berseem	Rabi	15.5	67	0.43

Table 30. Nutrient content of Green manures

Green manure	% moisture	N	P_2O_5	K_2O
Sesbania/Dhaincha	78	3.5	0.6	1.2
Sunnhemp	75	2.3	0.5	1.8
Cowpea	80	2.3	0.5	2.1
Guar	75	3.0	0.4	1.6
Pongamia	75	3.3	0.4	2.4
Glyricidia	80	2.9	0.5	2.8
Calatropis gigantea	75	0.42	0.12	0.37
Wild indigo	75	3.2	0.3	1.3
S.rostrata	78	3.1	0.3	1.46

In situ green manure systems

1. *In situ* green manure
2. Annual green manures are grown in small area and used for green manuring in another site of 3-4 times the area of the green manure cropped area. Roots remain in the first field add to fertility.
3. Green manure crop is harvested and used for compost production.
4. Interplanting of green manure crops (*Sesbania cannabina*) in rice for providing green manure to the succeeding crops.

Strategies to mitigate loss of growing season under *in situ* green manure

One of the major limitations of *in situ* green manures is the loss of one season and additional cost of cultivation. These limitations can be mitigated by adopting following techniques

- **Improved fallow:** Replacing natural fallow vegetation with green manure crops to speed up regeneration or soil fertility.
- **Alley cropping or simultaneous fallow:** Fast growing shrubs or grasses are planted in rows and are regularly pruned; prunings are used as mulch/incorporated. *Faidherbia albida*, glyricidia, subabul.

- **Integrated tree farming:** Integration of trees with crops *e.g.*, *Erythrina poeppigiana* in coffee plantations. Trees are grown in coffee, rubber and tea plantations to provide shade. The loppings or prunings are used as green manure.

- **Relay fallowing:** Bush legumes are sown in the food crops. The legume biomass is used as mulch or manure in the dry season *e.g.*, *Sesbania rostrata*, *Tephrosia purpurea* is grown along with rice during summer which is used as green manure for July transplanted rice. Other green manure crops suitable are *Sesbania speciosa* and *Cassia leschenalltiana.*

- **Live mulch:** In which rows of food crops are sown into low but dense cover crop of grass or legumes e.g. *Centrosema pubescens, Pueraria phaseoloides,* cowpea and *Arachis prostrate.* The dense cover crops act as live mulch. The rate of transpiration of live mulch crop should be lower. These live mulch crops are sown in orchards or plantations.

- **Sole cropping of legumes:** Legumes grown in sequential cropping with food crops can be harvested for grain/vegetable and residues are incorporated into the soil for the succeeding crop.

  Greengram – sorghum

  Cowpea – finger millet

  Greengram – fingermillet

  Sunnhemp – wheat

  Blackgram – wheat

  Sustained benefits possible if the system is followed for 3-5 years.

  Pigeon pea provides good ground cover, fixes atmospheric nitrogen. roots penetrate deep into the soil. Pigeon pea may be grown for two to three years followed by grain crop for another two to three years will greatly increase the land utilization index.

- **Green manure as main crop in rotation with food or commercial crops**

  Rice – sunnhemp

  Rice – berseem (*Trifolium alexandrinum*)

  Wheat – groundnut

  Maize- green manure- cotton

  Cotton – sunnhemp

- **Intercropping shade tolerant legumes:** In tall growing food crops. Cowpea or *Calapogonium muccunoides* is interplanted with maize.

- **Interplanting of legume with grasses** (grass legume mixture): Green manure crops cover the ground and improve the soil moisture, soil temperature and soil structure.

- **Inter row sown crops:** Quick growing green manures are grown along with the main crop

  Rice + Dhaincha

  Castor + cowpea, six weeks after sowing, cowpea is incorporated into the soil.

  Sorghum + Cowpea

  Sunflower + Greengram/blackgram

  Pigeonpea + groundnut

  Sugarcane + Greengram/blackgram

  Cotton/maize + Sunhemp/cowpea/horsegram

  Cotton or sugarcane + Berseem

  In drilled rice, seeds of sunnhemp are mixed with rice at 2–3 kg per ha. After about two months sunnhemp is incorporated into the soil. About 3000 kg green matter may be added per ha.

  Sugarcane + sunnhemp inter cropping: incorporation of sunnhemp tops or entire plant increased the cane yield to 155 t/ha from 128 t per ha in control.

- **Catch crops:** Quick growing green manure crops like sunnhemp, dhaincha, pillipesara are grown during May-June and incorporated into the soil during July–August.

- **Rabi cropping areas:** In vertisols where *rabi* is the major cropping season green manure crop is grown during the kharif.

- **Harvest for vegetable:** Cowpea or green gram pods may be harvested for vegetable and the haulm may be incorporated.

Transplanting technique of Sesbanea: Seeds are sown in the nursery at the rate of 70–80 kg per ha. One ha of nursery produces sufficient seedlings to green manure 50 ha of rice fields. Superphosphate at 225 kg per ha and compost at the rate of 7.5 t/per ha is applied in nursery plot. When seedlings are about 7–10 cm taller they are transplanted in the rice field to be green manured at a row spacing of 2–3 m and intrarow spacing of 30–45 cm. The *Sesbania* plants are topped when they attain a height 30–150 cm to stimulate branching and leaf production. A full grown plant attains a height of 3 m and weighs about 4 kg. Twenty days after transplanting *Sesbania*, superphosphate at 75 kg/ha and compost at 2.25 t ha is added. Fifteen days after harvesting rice ammonium sulphate at 50–60 kg/ha is added and ploughed under 3 days before transplanting of succeeding rice crop. The yield of green manure is 15–22.5 t/ha containing 82.5 kg N, 10–16 kg P_2O_5 and 23–34 kg K_2O.

Sesbania speciosa: There are four ways of green manuring

1. Three to five days prior to harvest of rice crop, *Sesbania* seeds at 50 kg per ha are broadcasted.
2. Seeds are broadcasted into the ploughed land at 35–50 kg per ha after harvest of preceding rice crop.
3. Three weeks old seedlings of *Sesbania* can be grown along the borders of the field bunds during the first cropping season and utilized as green manure for the second crop.
4. Border planting of *Sesbania* at spacing of 5–10 cm in one ha will yield about 5000–8000 kg. At the time of rising rice nursery for the first crop 0.75 kg of seeds of *Sesbania* may be sown in 100 sq m of nursery area. At transplanting the rice seedlings *Sesbania* seedlings are also transplanted along the border.

Improving the efficiency of green manure plants

1. **Liming:** application of lime in acidic soils.
2. **Phosphate fertilization:** improves the growth of rhizobia, root growth, inorganic phosphorus taken up by the green manure crop will be transformed into organic form. Single super phosphate at 40 kg/per ha may be applied in neutral and alkaline soils. In medium acidic soils, bone meal, and basic slag and in strongly acidic soils rockphosphate may be used.
3. Nitrogen and potash are also added in soils poor in nitrogen and potassium.
4. Addition of molybdenum and boron will improve the growth of green manure crops
5. Inoculation with appropriate nitrogen fixing culture. Treatment with mycorrhizal cultures is also beneficial.

Green leaf manure

Glyricidia sp.

Glyricidia maculate on bigger bunds can produce 12 kg per tree for July crop and 6 kg per tree for Janurary crop.

Glyricidia sepium (Jacq.) Steudel belongs to sub-family papilionoideae and family leguminosae. It is a multipurpose shrub performs well in saline vertisols. It is used in reclamation of fly ash slurry pits and open cast mining sites. It does not tolerate frost and cold. It can be used in alley cropping or alley farming.

Glyricidia maculate H.B.K. native of Mexico and *Glyricidia Brenningii*: native of America. These two are lesser known species.

Table 31. Nutrient contribution (kg) from *Glyricidia sepium* pruning's from alley cropping

Nutrient	Amount of pruning's in alley cropping kg ha⁻¹cutting⁻¹			
	500 plants	1000 plants	2000 plants	3000 plants
Nitrogen	15–23	30–45	60–90	90–135
Phosphorus	1–1.5	2–3	4–6	6–9
Potassium	8–18	16–36	32–72	48–108
Calcium	7	14	28	42
Magnesium	2–3	4–6	8–12	12–18

Nutrient content of *Glyricidia sepium* is 3.0–4.5% N, 0.2–0.3% P, 1.6–3.6% K, 1.4% Ca and 0.4–0.6% Mg.

Leucaena Sp.

Leucaena leucocephala (Lam.) de Wit: Leaves contain 3% N and decompose fast. *Leucaena* when planted as alleys with rabi sorghum and loppings are used as manure add 87.6 kg N per ha. One tonne of dry leaves add 30–35 kg N, 2.7 kg P, 14 kg K, 8.0 kg Ca, and 1.4 kg S.

Sesbania sp.

Sesbania grandiflora (L.) Poir (Agati/Avise): Nitrogen content is 5% produce 20 t dry matter per ha.per year.

Sesbania bispinosa is an annual shrub or short lived perennial that can grow to 6–7 m tall. It is also known as *Sesbania aculeate* (Willd) Poir and as *Sesbania cannabina* (Retz.) Poir and by the common name Dhaincha. It is a Multipurpose shrub. It grows under tropical, subtropical and semi-arid regions. It is resistant to drought, soil acidity, alkalinity and water logging.

Sesbania sesban (L) Merril (Sesban) (= *S. aegyptica* Poir) fast growing, can be grown in tropics, tolerant to salinity and floods.

Aquatic green leaf manure plants

Water bodies (waste water lagoons, industrial waste water, water courses) can be used to grow aquatic plants for green biomass production.

Table 32. Aquatic Green manure production and nitrogen yield.

S. No.	Aquatic plants	Green biomass yield (t/ha)	N/ha	Duration
1.	*Alternanthera philoxorides*	225–375	344–572	1 yr
2.	*Eichornia crassipes*	87.5	122	1 yr
3.	*Eichornia crassipes* (grown on sewage water)	250	348	1yr
4.	Azolla	200–300	547–821	1 yr

C/N ratio: it is one of the factors deciding the nutrient release for the crop. In neutral and alkaline soils of temperate regions a C: N of < 22 is required, in acid soils it is approximately 15. Leaf portion of the green manure is rich in nitrogen decompose quickly (4–5 days). Stem and woody portions decompose slowly and act as compost or bulky organic manure. Loppings of Glyricidia has lowest C: N (5–6.7) act as organic fertilizer. Tender material may incorporated just before planting while woody materials require longer period. Twelve week old dhaincha green manure incorporated require 8 weeks for decomposition while 8 weeks dhaincha decomposes immediately.

Table 33. C: N of different Green manures

Green manure	Carbon	Nitrogen	C/N
Sesbania aculeate	37.1	2.79	13.2
Crotalaria juncea	40.6	3.02	13.5
C. brownie	32.8	4.96	6.6
S.speciosa	44.3	2.32	14.7
C. striata	30.5	3.94	7.8
Phaseolus aureus	23.8	3.18	7.5
Cassia tora	36.0	2.48	14.5
Indigofera anil	31.1	3.35	9.3
Tephrosia purpurea	35.1	3.44	11.6
Aeschynomene americana	35.1	3.33	9.5
Vigna sinensis	27.5	3.19	6.6

Economics of raising a legume especially for biological nitrogen fixation is not attractive. Intercropping of legumes with cereals and supplying starter nitrogen dose under optimum supplies of other yield limiting nutrients is sustainable. Green manuring *per se* does not significantly increase organic carbon in tropical soils but benefits accrue from the substitution of chemically fixed nitrogen and enhanced biological activity. Biochemical process of mineralization, mobilization and transformation and transport of the nutrients are the added advantage.

Crop Residues

Crop residues are defined as the noneconomic plant parts that are left in the field after harvest and are discarded during crop processing.

Major crop residues in India come from sugarcane, coconut, wheat, paddy, maize, sunflower, cotton, redgram, castor, small grains and grasses. Management of sugarcane trash, coir pith, bagasse of sugarcane, gin trash of cotton as a source of plant nutrients and soil amendment assumes greater importance.

- They are available on farm, does not involve transportation cost.
- Does not involve loss of a cropping season as in case of green manures.

Limitations in use of crop residues

- They do not decompose readily due wider C: N ratio.
- Decomposition of crop residue requires adequate soil moisture.
- They require size reduction for faster decomposition which involves handling of bulky material.
- If the residues contain the inoculum of pest or disease causing organism, the succeeding crop is likely to be infected.

Management of crop residues

- Crop residues mulched on the soil surface conserve soil moisture, control weeds and moderate soil temperature.
- Conservation tillage is one of the management techniques of crop residues where in 30 per cent of the residues are left on the soil surface.
- Selective removal of the residues should be practiced.
- Substituting high quality fodder for crop residues as animal feed, alley cropping, using wastelands for forage and pasture production.
- Burning of residues in the field should be avoided.
- Combined use of crop residues with fertilizers should be practiced to hasten the residue decomposition and reduce the nitrogen depression or deficiency in the crop.
- Crop residues may be composted either *in situ* or *ex situ* adopting improved composting techniques *viz.* coir pith composting and sugarcane trash composting.

Coir pith composting: Heap method of composting is followed. Hundred kg of coir pith is spread in an area of 5m x 3m. The material was inoculated with one bottle spawn of *pleurotus sajor caju*. Another 100kg of coir pith is spread over the first layer. Over this one kg of urea is sprinkled. The alternate layers of coir pith – *pleurotus*, coir pith – urea are repeated for one tonne of coir pith. Five bottles of pleurotus sp. and 5 kg of urea are required. Water is sprinkled to maintain the moisture content of the material > 60 per cent. At the end of 30days coir pith turns to black mass with considerable reduction in C: N ratio.

Table 34. Nutrient composition of raw and composted coir pith

Composition	Raw coir pith	Composted coir pith
N (%)	0.260	01.18
P (%)	0.025	00.11

(Contd.)

Composition	Raw coir pith	Composted coir pith
K (%)	0.750	01.35
OC (%)	32.400	23.30
C/N	124	20

Sugarcane trash composting

The trash is spread in an area of 5m x 3m to a height 15cm. Over this layer pressmud is spread uniformly to a thickness of 5cm. Rockphosphate, gypsum and urea in 5: 4: 1 is mixed and applied at the rate of 10kg/100kg of sugarcane trash. Moisture content of the material is maintained above 40 per cent. The alternate layers of sugarcane trash, pressmud, mixture of gypsum, rockphosphate and urea are repeated upto a height of 1m to 1.5m. At the surface the heap is covered with cattle dung paste. Water sprinkled once in 15days. After three months thorough turning is given, and heaped again. One more turning is given after a month. Outside material is mixed with that from inside. Compost will be ready in about five months. N, P_2O_5 and K_2O concentration are in the ratio of 2.73:1.81: 1.31.

QUESTIONS

Answer the following

1. Classify the greenmanures.
2. What are the characteristics of *in situ* green manure plants?
3. What are the strategies for coping loss of growing season of *in situ* green manuring?
4. How do you improve the efficiency of *in situ* green manuring?
5. What are the benefits of in situ green manuring?
6. State the importance of crop residues.
7. What are the limitations of crop residue use in crop production?
8. Explain crop residue management techniques.
9. Explain coir pith composting.
10. Explain sugarcane trash composting.

CHOOSE THE CORRECT ANSWER

1. Green manure suitable for water logged situation is
   (a) Dhaincha
   (b) Sunnhemp
   (c) Cluster bean
   (d) Pillipesara

2. Green biomass yield of sunnhemp is
   (a) 12 t/per ha   (b) 15 t/per ha
   (c) 20 t/per ha   (d) 30 t/per ha

3. Aquatic greenmanure plant
   (a) *Eichornia crassipus*   (b) *Sesbania rostrata*
   (c) *Trifolium alexandrium*   (d) *Calatropis gigantea*

4. The coir pith decomposing culture is
   (a) *Trichoderma viridae*   (b) *Pleurotus sajur caju*
   (c) *Pseudomonas striata*   (d) *Aspergillus awamorii*

5. Stem nodulating green manure is
   (a) *Cyamopsis tetragonaloba*   (b) *Crotalaria juncea*
   (c) *Sesbanea rostrata*   (d) *Ponagamia glabra*

6. The C: N of coir pith is
   (a) 40:1   (b) 124:1
   (c) 80:1   (d) 20:1

Chapter 11

Biofertilizers: Role and Management

Biofertilizers act as complementary and supplementary sources of plant nutrients. Biofertilizers help in increasing the biologically fixed atmospheric nitrogen and enhancing phosphorus availability to the crops. Among the biofertilizers for increasing nitrogen supply, nitrogen fixing bacteria (*Rhizobium, Azotobacter* and *Azospirillum*), blue green algae (BGA) and *Azolla* are important. The availability of phosphorus is improved by phosphorus solublizing microorganisms (PSM) and mycorrhizae (VAM).

Rhizobium is of vital importance to various pulses and some of the oilseeds and fodder legumes as 80–90 per cent of their nitrogen requirement is met through biological nitrogen fixation (BNF). Rhizobium fixes atmospheric nitrogen ranging from 50–60 kg per ha in groundnut, 100–300 kg per ha in alfalfa. In pulses it is estimated at 50–110 kg/per ha. Field studies indicated a yield increase of 14–53 per cent due to *Rhizobium* inoculation. Nitrogen left over the demand of legumes in combination with the nitrogen present in roots of the legumes will enhance the soil fertility. Consequently the nutrient needs of the subsequent crop from applications will be reduced.

Table 35. Estimated biological nitrogen fixed in legumes.

S. No.	Crop	Fertilizer nitrogen equivalents (Kg N per ha)	
		Nitrogen fixed (kg N/ha/year)	Residual effect in succeeding crop
1.	Alfalfa (*Medicago sativa*)	100–300	
2.	Clover (*Trifolium spp.*)	100–150	83
3.	Chickpea (*Cicer arietinum*)	26–63	60–70
4.	Cowpea	53–85	60
5.	Green pea	50–55	30
6.	Groundnut	112–152	60
7.	Guar	37–196	
8.	Lentil (*Lens culinaris*)	35–100	18–30
9.	Pea (*Pisum sativum*)	46	20–32
10.	Pigeon pea	68–200	20–49
11.	Soybean	49–130	

Rhizobium consists of three species

Rhizobium leguminosarum with three var. *viz. trifolii, phaseoli* and *viceae.*

Rhizobium loti

Rhizobium meliloti

The new genus *Bradirhizobium japonicum*

Rhizobia are also known to occur as endophyte in the roots of rice, wheat and maize. It is recognized that rhizobia are aquatic, epiphytic and endophytic in addition to soil bacteria. Many soils contain non symbiotic saprophytic rhizobia both in the soil and in the rhizosphere of legumes and other plants. Performance of crops to biofertlizer application is highly unpredictable because of their susceptibility to biotic and abiotic stresses. Effectiveness of the introduced strain depends on the competitiveness of the strain. The parameters of virulence of the introduced microbe are:

- Numerical superiority
- Rhizobial multiplication
- Survival
- Tolerance to antibiotic agents
- Chemotaxis
- Motility efficiency of attachment to roots
- Growth rate and
- Multiplication after root/nodule senescence.

Rhizobium has two forms

Fast grower: Rhizobium isolated from legumes of temperate region such as *Rhizobium trifoli, Rhizobium leguminosarum, Rhizobium phaseoli* and *Rhizobium meliloti* are designated as fast growers having generation time of less than six hours.

Slow growers: *R japonicum, R. lupini* and Rhizobia belonging to cowpea miscellany group isolated from legumes of tropical region have generation time more than six hours.

Certain group of leguminous plant shows specificity or preference to a single kind of Rhizobium. The bacteria that nodulate plants belonging to these groups are designated as 'species'. All plants nodulated by a Rhizobium species constitute a 'cross inoculation group.

Symbiosis found outside the established cross inoculation groups is incapable of fixing atmospheric nitrogen. Most of the cultivated legumes in India belong to cowpea cross inoculation group.

Table 36. Cross inoculation groups of Rhizobium

Rhizobium species	Cross inoculation group	Legumes
R. trifoli	Clover group	Trifolium
R meliloti	Alfalfa group	Melilotus, medicago
R. phaseoli	Bean group	Pheseolus
R. lupini	Lupine group	Lupinus, ornithopus
R. leguminosarum	Pea group	Pisum, vicia, lens
R. japonicum	Soybean group	Glycine
Rhizobium sp.(miscellany)	Cowpea	Vigna and Arachis

Nirtrogen fixation by Rhizobia is a high energy consuming process. Supply of photosynthates by the host plant to rhizobia is primary regulator for nitrogen fixation. It has been estimated that host plant has to lose 15–20 kg of drymatter per kg of nitrogen fixed. Nonetheless, legumes grown under low fertility without addition of fertilizer nitrogen can substantially contribute towards nitrogen economy. Further the fixed nitrogen enters the soil nitrogen pool as stable organic nitrogen and resistant to short term losses *via* leaching and denitrification.

In general high nitrogen content in soil, applied or residual, reduces nodulation and nitrogen fixation. To improve the contribution of biological nitrogen fixation under such circumstances, soil nitrogen must be managed by including an appropriate nitrate tolerant high nitrogen fixing legume or a genotype of a given legume and/or appropriate cropping and management practices. Based on soil nitrogen mineral status at sowing need based application at low doses of nitrogen for increased legume productivity and to maintain high soil fertility has been suggested. Fertilizer management in intercropped legume should be followed without affecting biological nitrogen fixation. Shading effect of the component crop can reduce biological nitrogen fixation. Strip cropping and selection of noncompeting non legume crop can overcome both these problems. Besides tillage, land treatments and application of the nutrients to enhance biological nitrogen fixation play significant role.

Factors affecting nitrogen fixation

Moisture: legume symbiosis is sensitive to both drought and waterlogging. During dry periods root hair formation is inhibited. Moisture stress inhibits nodule formation. Moisture in excess of 25% of field capacity and at least up to 40% moisture enhance nitrogen fixation activity in a vertisol.

Light: basic factor required for the host plant to synthesize adequate photosynthates.

Temperature: At higher temperature photosynthesis is drastically reduced and hence nitrogen fixation can be indirectly affected by reduced supply of photosynthates. Both higher and lower soil temperatures are harmful. Higher acetyl reduction assay was observed when roots are incubated at temperature 26 °C than at 20 °C or 30 °C. for most of the crops 24–30 °C is found optimum.

Nutritional factors: The elements specifically required for symbiotic nitrogen fixation are molybdenum, cobalt and iron. Soil acidity together with toxicity caused by aluminum or manganese (or both) and deficiency of phosphorus, sulphur, calcium and molybdenum are some of the factors limit the grain yield of tropical legumes. Nodulation is sensitive to acidity than plant growth. It can fix nitrogen in acid soils in the pH range of 4.5–5.5. Nodulation and nitrogenase activity depressed by nitrogen concentration >25 ppm NO_3– N. Adequate lime and phosphorus application is essential for higher nodulation.

Salinity: The rhizobia can tolerate a higher level of salinity than the host legume.

Insect damage: *Sitona sp.* and *Revelllia sp.* are known to attack legume nodules. Extensive damage to nodules was caused by *Revellia angulata*.

Cropping systems also influence Rhizobia population. Generally after puddled rice the Rhizobium population declined. The population of cowpea cross inoculation group increased in redgram and mung intercropping system

Effective Rhizobia strains: Some of the effective strains in different crops are as follows

Redgram: IHP 195 and IC 3100 developed at ICRISAT.
CC-1, A 2 and A 19 developed at Coimbatore, Pantanagar and IARI.

Bengalgram: F 75 for Northwestern and Northeastern plains zones.
H 45 for central zone.

IC 76 for peninsular zone.

Groundnut: NC 92

Methods of Rhizobia inoculation

Seeds are inoculated using 10% jaggery or sugar solution. At least 10^5 to 10^6 viable Rhizobia strains are required per seed.

Seed pelleting using any one of the materials *viz.*, calcium carbonate, dolamite, gypsum, bentonite, rock phosphate, super phosphate, talk powder, charcoal, peat or basic slag. Pelleting is done using the stickers *viz.*, methoxy and ethoxy substituted gums and gum arabica. More inoculum may be treated. It provides better environment for Rhizobia.

Soil inoculation is useful when seeds are treated with chemicals or when inoculation was not done at sowing. Granular peat inoculum is drilled or placed in seed rows or applied within 15–20 days after sowing.

Compatibility with plant protection chemicals:

When seeds are treated with thiram or captan at 2.5–3.0 g/per kg seed, use double the recommended rate of inoculum. Ceresan is not compatible with Rhizobia.

Asymbiotic nitrogen fixers

As against the high amount of carbon energy utilized by legume-rhizobium symbiosis large amount of nitrogen is fixed in the rhizosphere of non-legume plants by diazotrophs is called associative nitrogen fixation or rhizocoenoses which has increased the interest in biological nitrogen fixation.

Azotobacter chroococcum is one of the dominant non-symbiotic nitrogen fixing heterotrophic bacterium. *A. Chroococcum* in Indian soils rarely exceeds 10^4–10^5 per g of soil. *Azotobacter* is dominant in the rhizosphere of crop plants except for wheat, which harbour anaerobic clostridia like organisms. Certain amino acids, sugars, organic acids and vitamins in root exudates of crop plants serve as energy source.

The roots of the transplanted crops are dipped in slurry of azotobacter before planting. The seeds are mixed with carrier-based cultures and dried in shade. Inoculation may be done after crop establishment by pouring the inoculum slurry near the root zone.

Apart from fixing atmospheric nitrogen, biologically active substances *viz.* indole acetic acid, gibberallins, B- vitamins, nicotinic acid, pantothenic acid, biotin and heteroauxin are produced. It has ability to produce antifungal antibiotics and fungistic compounds against pathogens like fusarium, alternaria and trichoderma.

Significant increase in yield was seen in rice, cabbage, brinjal, sugarcane, cotton. About 30 to 35kg nitrogen per ha is supplied by Azotobacter.

Azospirillum: Beijerinck in 1925 described *Spirillum lipoferum* as a nitrogen fixing bacteria. Sen (1929) made one of the earliest suggestions that nitrogen requirement of cereal crop can be met by the activity of associated nitrogen fixing bacteria such as *Azospirillum.* More than 30grass species are associated with rhizosphere nitrogen fixing bacteria. C_4 plants colonized by *Azospirillum lipoferum* and C_3 plants by *Azosprillum brasilense.* Recently two more species *viz. A. amazonense* and *A. seropedica* were identified. Associative nitrogen symbiosis is generally known to be active in graminacious crops *viz.* sorghum and millets. Genotypic variation in sorghum and millets in their nitrogenase activity was observed. Azospirillum survives in the media of farmyard manure + soil, farmyard manure +charcoal or farmyard manure alone for 31 weeks.

Azospirillum besides fixing nitrogen produces several growth promoting substances. In sorghum, maize, pearlmillet and setaria about 25% N requirement may be substituted by Azospirillum inoculation. Eleven per cent increase in grain yield has been reported in rice, wheat, sorghum, maize and pearlmillet.

Methods of *Azospirillum* inoculum:

- Seed coating similar to *Rhizobium.*
- **Root dipping:** 600 g culture mixed with sufficient quantity of water and roots are dipped for 15–30 min.
- **Direct seeded rice:** seeds mixed with inoculums using rice gravel for sticking.
- Two kg inoculum mixed with 25 kg farmyard manure and 25 kg soil applied in dryland crops.
- Five kg inoculum mixed with 500 kg of powdered farmyard manure applied in 2 splits at 30 and 60 days after planting at the base of the clumps in sugarcane.

Acetobacter diazotrophicus is a saccharophilic bacterium associated with sugarcane, sweet potato and sweet sorghum. These bacteria can be multiplied in nitrogen free medium and mixed with the carrier material and used for inoculating sugarcane.

Azolla pinnata var. *pinnata* a fern consisting of small, overlapping bilobed leaves. The algal symbiont *Anabaena azollae* is associated with the dorsal leaf lobes from the onset of their development and is never in direct contact with the external environment. Growth and development of both azolla and anabaena are synchronous. The heterocyst is the site of nitrogen fixation and occurrence of reduced condition in it was confirmed. The ammonia released from the nitrogen fixation of the symbiont is released into the leaf cavity, which is absorbed by fronds of ammonium assimilating enzymes. The hair cells lining the leaf cavity are suggested as the site of metabolic exchange between azolla and its symbiont.

Azolla culture production

The field divided into plots of 20 m × 2 m by providing suitable bunds and irrigation channels. Water is maintained at a depth of 10cm. Ten kg fresh cattle dung mixed with 20 lit of water is sprinkled in each plot and Azolla inoculum at 8 kg is introduced to each plot. Superphosphate (100 g) is applied in 3 splits at 4 days interval as top dressing. For insect control furadon granules at 100 g per plot may be applied at 7 days after inoculation. Azolla culture is harvested after three weeks and introduced into the main rice fields. About 4 plots are required to provide inoculum required for 1 ha. From each plot 40–50 kg fresh azolla may be harvested.

Nutritional value of azolla: Azolla contains 3–5 per cent nitrogen on dry weight basis and 0.2–0.4 per cent on fresh weight basis. Azolla turns into compost within 1–2 weeks, it contains about 5 per cent nitrogen on dry weight basis. Phosphorus and potassium content is 0.5–0.9 per cent and 2.0–4.5 per cent respectively.

Pyralis and snails are the major pests; black rot is the major disease.

A thick mat of azolla can supply 25–30 kg each nitrogen and potassium respectively.

Blue green algae: These are the photosynthetic organisms owing to the presence of nitrogen fixing cells (heterocysts). It uses sunlight as energy source and water as source of reductant for photosynthesis and nitrogen fixation. The predominant genera of blue green algae in Indian soils are *Anabaena, Nostoc, Aulosira* and *Calothrix* and *Tolypothrix*. They are filamentous with chain of vegetative cells including specialized cells called heterocyst which function as a sort of micronodule. Blue green algae is applied at 10 kg per ha at 10 days after planting. The field is kept water logged for a couple of days immediately after application. Blue green algae contributes 20–30 kg nitrogen per ha. Besides it provides auxins and amino acids. Improves aeration and reduces sulphide injury in rice fields.

Production of blue green algae inoculum: 20 m × 1 m × 22 cm polythene lined plots are prepared. The pH of water is neutralized to 7. Starter culture is inoculated to plots. A thick algal mat is formed in 15days. This was allowed to dry. The flakes are collected and dried. About 10 kg inoculum may be produced from a plot of 20 m × 1 m.

Phosphorus solubilizers

Phosphorus in ionic form ($H_2PO_4^-$, HPO_4^{--}) at the root cell plasma membrane surface will be effective as a plant mineral nutrient. The amount of phosphorus retained in various mineral complexes is a function of soil pH, soil type, moisture content *etc.* Microbial agent that converts the fixed forms of phosphorus $Al(OH)_2 H_2PO_4$, $Fe(OH)_2 H_2PO_4$, $Ca_{10}(PO_4)_6 Fe_2$ into available ionic forms ($H_2PO_4^-$, HPO_4^{--}) that can be taken up by plants. These organisms have been referred to as mineral phosphate solubilizers.

Bacteria: *Agrobacterium radiobacter, Bacillus polymyxa* and *Bacillus megatherium*

Fungi: *Aspergillus awamorii, Aspergillus flavus* and *Aspergillus niger*

Mycorrhizae

Symbiotic association of fungus and plant root system is of two types *viz.* ectomycorrhizas and endomycorrhizas and the former colonize the woody trees and forest species. Our interest is in endomycorrhizal association of annual plants. These fungi are phycomycetes of the family endogonaceae. The most important group of endomycorrhizae is Vesicular - arbuscular mycorrhizae (VAM). They form vesicles (bladder like structures) and arbuscles (shrub like structures) within the root cortex, hence the name. The hyphae enter the roots either through epidermis or root hairs and extend out as far as 2 cm–8 cm from the root surface thus accomplish the extension of root geometry. Diameter of AMF hyphae is 3–4 µm as against the root diameter of > 10 µm and can make contact with soil particle and/or explore pores and cavities that roots would not otherwise contact. It also decreases the toxicity of iron, boron and aluminum toxicity in acid soils by inhibiting the acquisition of toxic minerals. The root or hyphae exudations may decrease the reductions in the rhizosphere like manganese. VAM has been found in cereals, grasses, legumes, cotton, tobacco, potato, sugarcane but the exception being the cruciferae and chenopodaceae. The symbiotic association between the fungus and the host plant is that the latter provide carbohydrates and the former makes nutrients more available. VAM is of considerable significance for nutrients that are diffusion limited such as phosphorus, zinc, copper and iron. Enhanced nutrient uptake is essentially due to extensive hyphae covering the soil volume. *Glomus fasciculatum*, *Glomus mossae* and *Glomus tenius* are suitable for pulses, oilseeds and vegetable crops.

VAM is an obligate symbiont it does not complete the life cycle in the absence of a suitable host plant. Resting spores are difficult to remain in isolation from the mycelia and the chlamydospores.

Mycorrhiza culture production technique

A pot filled with soil and sand (1:1) both soil and pot was sterilized using formalin under polythene cover for a night. Next morning the cover was removed. The pH of the soil should be brought near neutral by adding lime. The starter mother culture of VAM (*Glomus fasciculatum*) collected from soil was inoculated 2 cm below the soil surface in the pot and then covered with the soil. The seeds of guinea (*Panicum maximum)* grass were sown in soil and optimum soil moisture is maintained. Maize or sorghum can be used as host plant. The plants are thinned to maintain the optimum number per pot size. Soil moisture was maintained at 70 per cent. The plants were allowed to grow for 60 days. At flowering the above ground portion is cut and the rhizosphere soil and the root materials together were collected. They were mixed, sieved in 2 mm sieve and then the roots were cut into small pieces. The root fragments were mixed with the sieved soil and preserved. The soil so collected contains mycorrhizal root fragments, hyphae and spores either attached to the root or free in the rhizosphere soil. The spore load 10–14 per g of soil as against the initial number of seven spores per g of soil.

QUESTIONS

Define

(a) Asymbiotic nitrogen fixers

(b) Biofertilizers

(c) Cross inoculation group

(d) Phosphorus solubilizers

(e) Acetobactor

Answer the following

1. What are the types of root nodulating rhizobium?
2. Nitrogen fixation is a high energy consuming process. Explain.
3. Explain the factors affecting nitrogen fixation.
4. Explain different methods of rhizobium inoculation.
5. What is rhizocoenoses?
6. Explain the different methods of Azospirillum inoculum.
7. Explain azolla culture production.
8. What are vesicles and arbuscles?
9. Describe the mechanism involved in nutrient mobilization by mycorrhizae.
10. Explain VAM culture production.

CHOOSE THE CORRECT ANSWER

1. The nitrogen content of azolla on dry weight basis
   (a) 3–5%   (b) 7–8%
   (c) 1–2%   (d) 5–6%
2. The rhizobium nitrogen fixation in groundnut is estimated at
   (a) 112–152 kg per ha   (b) 50–100 kg per ha
   (c) 25–30 kg per ha   (d) 150–200kg per ha
3. The yield increase due to Azospirillum sp..inoculation is
   (a) 11%   (b) 5%
   (c) 20%   (d) 30%
4. Nitrogen contribution due to blue green algae inoculation in paddy is
   (a) 10–20 kg per ha   (b) 30–40 kg per ha
   (c) 5–10 kg per ha   (d) 20–30 kg per ha

5. The nitrogen fixed by Azotobactor is
   (a) 15–20 kg per ha   (b) 35–45 kg per ha
   (c) 30–35 kg per ha   (d) 10–15 kg per ha

6. The mycorrhizal hyphae extends the root by
   (a) 2–8 cm   (b) 2–8 mm
   (c) 2–3 cm   (d) 2–3 mm

7. The diameter of the VAM hyphae is
   (a) 3–4μm   (b) 3–4 mm
   (c) 1–2 cm   (d) 3–4 cm

8. The host plant for VAM production is
   (a) *Panicum maximum*   (b) *Triticum aestivum*
   (c) *Oryza sativa*   (d) *Vigna sinensis*

9. Heterocysts are found in
   (a) *Azatobacter*   (b) Blue green algae
   (c) *Acetobacter*   (d) *Azospirrilum*

10. Root nodules are damaged by the pest
    (a) Piralis sp.   (b) Revellia sp.
    (c) Nematodes   (d) Snails

Chapter 12

Integrated Nutrient Management in Cropping Systems

Integrated plant nutrition system conceptualized by food and agriculture organization is the maintenance or adjustment of soil fertility and of plant nutrient supply to an optimum level for sustaining the desired productivity through optimization of benefits from all possible sources of plant nutrients *viz.* organic manures, fertilizers, crop residues, compost or nitrogen fixing crops *etc.* in an integrated manner.

OBJECTIVES OF INTEGRATED NUTRIENT MANAGEMENT

1. To increase the availability of nutrients from all sources in the soil during the growing season.
2. To match the demand of nutrients by the crop and supply of nutrients from all sources through the labile soil pool both in space (rooting zone) and time (growing season).
3. To optimize the functioning of the soil biosphere decomposition, control of pathogenic organisms by their natural enemies, biological formation of soil structure, decomposition of phytotoxic compounds in soil.
4. To minimize the loss of nutrients to environment *e.g.*, through NH_3 volatilization and denitrification in the case of nitrogen, surface runoff and leaching of nutrients beyond rooting zone.

Steps in Integrated Nutrient Management

1. Obtain accurate soil information for each field or management unit. This could require new soil map or adaptation of existing maps to find out soil nutrient status.
2. Yield potential for each field should be estimated based on the productivity of soils and intended management practices. Take the average of past 5–7 years.
3. Calculate plant nutrients required to achieve the yield potential. Nutrient uptake (use by the crop) and nutrient removal (physical displacement of nutrients from the field in the harvest) data available for different crops. Uptake and removal of nutrients are not the same.

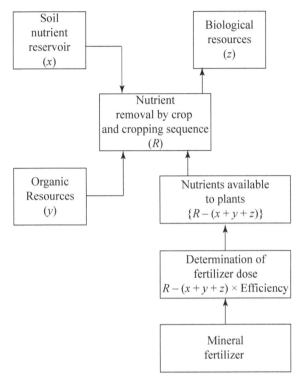

Illustration of the integrated plant nutrition system concept

4. Determine the plant available nutrients in manures or other organic amendments. Accurate nutrient content of the manure is site/animal/diet/ management specific.
5. Estimate the residual nutrient contribution from fertilizers or manures applied in previous seasons. Manures are originally slow release nitrogen fertilizers. Usually about 50% of nitrogen content is available to growing plants the first year following application. Subsequent use is usually sliding scale.
6. Apply animal manure and/or commercial fertilizers to supply nutrients when needed by the growing crops using best management practices.

IPNS in dry lands

Special characteristics of drylands

The soil resource provides a poor nutrient supply base and its erosion and degradation take place rapidly resulting in lower soil productivity.

Biomass availability in dry lands is of a lower order. Many times there is a conflict among various objectives of nutrient, fuel and fodder applies.

Choice of bioinoculant resources is limited due to water deficient situations. For example it is not feasible to supply blue green algae and Azolla in many situations. *Rhizobia, Azospirillum* and *Azotobacter* may not be efficient.

Utilization of mineral fertilizer resources by plants is limited due to inadequate soil water.

Steps to develop a model for IPNS in drylands
1. Identification of suitable and potential cropping systems for the area.
2. Assessment of nutrient removal by the cropping systems for the desired yield levels.
3. Assessment of soil fertility status and its contribution towards nutrient supply.
4. Establishment of fertilizer rates on the basis of total nutrient requirements for the cropping system and the availability of nutrients through soil, organic and biological resources while taking into account fertilizer efficiency.
5. Evaluation and monitoring of the effect of the above mentioned system on the long-term crop yields, soil productivity and ecology.

SELECTED INTEGRATED NUTRIENT MANAGEMENT MODELS

Irrigated situation

Food and agricultural organization developed integrated plant nutrient system (IPNS) models for selected cropping systems. Fertilizer recommendation for rice-wheat and maize-wheat is 240:90:120 kg N: P_2O_5: K_2O per ha with an expected grain yield under optimum management conditions is 11 t/per ha. By only inclusion of a short duration grain legume *e.g.* cowpea in the system and incorporating its residues into the soil after harvesting grain can save 30 kg N per ha in addition to providing 1 t grain legume per ha. Supplementing the kharif crop with 5 t farmyard manure per ha in addition to growing of cowpea as stated above further reduced the fertilizer requirement by 70 kg N, 60 kg P_2O_5 and (60 kg) K_2O per ha in the system and still producing 12 t grain per ha. Raising a leguminous green manure crop and incorporating before rice transplanting can lead to a saving of 60 kg N per ha.

Table 37. Integrated plant nutrient system in dryland village in Solapur region for pearl millet – bengalgram system

Crop	Production	Nutrient removal (kg per ha)			Resources	Needs met through various nutrient sources (kg per ha)		
		N	P	K		N	P	K
Pearlmillet	1500	49.56	13.5	52.6	Soil source	60 (10)	5.3 (50)	149.4 (15)
					Organic source	13 (30)	2.2 (65)	5.4 (75)
					Biological resources (*Azospirillum* inoculation)	20	0	0
					Fertilizer	50 (40)	46.2 (20)	33.2 (80)
Bengalgram	1000	44.8	3.4	31.5	Soil resource	65 (20)	6.6 (20)	166 (40)
					Organic resource	13(25)	2.2 (50)	5.4 (65)
					Biological resources (*Rhizobium* inoculation)	20	0	0
					Fertilizer	15 (60)	6.6 (15)	0 (75)

Figures in () indicate per cent nutrient efficiency/release rate.

Table 38. Nutrient availability through various organic sources with the farmer

Type	N	P	K
Cattle	9.2	1.32	5.49
Buffaloes	2.1	0.31	1.25
Goat	1.5	0.18	0.17
Crop residues	5.7	1.23	2.32
Human	7.5	1.32	1.66
Total	26	4.36	10.79

Assumed level of utilization 50% for dung and 75% for urine for cattle and buffaloes, 80% for dung and 75% for urine for goat, 75% for crop residues, 80% faeces and 75% urine for human.

Factors of Fertilizer Management in Cropping Systems

Soil supplying power: soil contributions to the crops should be known before deciding on the quantum of fertilizer application. The results of long term fertility trials revealed that there is no appreciable change in the soil physical properties and soil deficiency for micronutrients as a consequence of multiple cropping. However, zinc in light alluvial soils (sandy soil) of Ludhiana and iron in medium black soils of Coimbatore have been found to be critical. The soil nutrient status estimated by soil analysis at the beginning of the season is altered by growing different crops during different seasons. The soil supplying power increases with legume in rotation, fertilizer application and addition of crop residues. The available nitrogen and potassium in soil after a groundnut crop are higher compared to initial status of the soil. But after pearl millet, only potassium status in the soil is improved and there is no change in phosphorus status.

Nutrient uptake by the crops: The total amount of nutrients taken by the crops in one sequence gives an indication of the fertilizer requirement of the system. Balance sheet approach is followed to know whether the amount of fertilizer applied is equal, more or less to the total uptake of nutrients by different crops in the system. The balance is obtained by subtracting the fertilizer applied to crops in the system from the nutrients taken up by the crop.

Residual effect of fertilizers: The extent of residues left over is the soil depends on the type of fertilizer used. Because of their mobility and solubility, nitrogenous fertilizers leave no residues after the crop is harvested. ^{15}N studies have shown that only 1-2 per cent of nitrogen applied to maize was taken up by the following wheat crop. However residues of nitrogen occur only when previous crop yields are poor. Phosphorus fertilizers and farmyard manure leave considerable residue in the soil which is useful for subsequent crops. Farmyard manure applied to the previous crops used only 50 per cent of its nutrients and rest was available for subsequent crops. The residues left by potassium fertilizers are marginal.

Legume effect: Legumes add nitrogen to the soil in the range of 15–120 kg N ha^{-1}. The amount of nitrogen added depends on the crop and also on the purpose for which it is grown. Greengram grown for grain contributes 15–20 kg N ha^{-1} to the succeeding crop. Cowpea grown for grain and fodder contributes 24 and 30 kg N ha^{-1} to the succeeding crop. Inclusion of leguminous green manures in the system adds 40–120 kg N ha^{-1}. The availability of phosphorus is also increased by incorporation of green manure crops. Potassium availability to subsequent crop is also increased by groundnut.

Crop residues: Add considerable quantity of nutrients to the soil. Cotton planted in finger millet stubbles benefits by 20–30 kg N ha^{-1} due to decomposition of finger millet stubbles. Deep rooted crops like cotton and redgram absorb phosphorus and other nutrients from deeper layers. Leaf fall and its subsequent decomposition add phosphorus to the top layers. Crop residues containing high C: N ratio like stubble of sorghum and pearl millet temporarily immobilize nitrogen. Residues of leguminous crops contain low C: N ratio and decompose quickly and release nutrients.

Efficiency of crops: Crops differ in their ability to extract and forage nutrients from different layers of soils and their capacity to utilize them for the production of economic products. Jute is more efficient crop for utilization of nitrogen followed by summer rice, rainy season rice, maize, potato and groundnut in that order. The order of phosphorus efficient crops are jute > summer rice > kharif rice > potato > groundnut > maize. Groundnut is more efficient in potassium utilization and the order of efficiency is maize > jute > summer rice > rainy-season rice > potato.

Fertilizer recommendation should be made to the cropping system considering all the above factors. For example, in wheat based cropping systems, inclusion of greemgram or blackgram needs 20–30 kg nitrogen less per ha. When phosphorus fertilizers were added to the green manure crop, there is no necessity to apply phosphorus to the succeeding wheat crop. In rice-wheat system, recommended dose of nitrogen of crops has to be applied. However, it is sufficient to apply phosphorus to wheat and potassium to rice but not for both crops of the system. In rice-rice –pulse cropping system, nitrogen has to be applied to both the rice crops, phosphorus to dry season rice and potassium, sulphur and zinc, if necessary, to the second crop. In rice based cropping system consisting of rice-rice in kharif and rabi and sorghum, maize, finger millet, soybean in summer, it is sufficient to apply phosphorus and potassium to summer crops only while nitrogen is applied to all the crops. Thus lot of fertilizer can be saved by following system approach in fertilizer recommendation. Sorghum following tobacco gives good yield. Any crop following sorghum is slightly affected. This may be nullified by inter-sowing indigo with sorghum. The indigo makes vigorous growth with occasional showers and gets ready for being ploughed into the soil in about 2 months after sorghum harvest.

In rice- wheat system application of 100% recommended NPK to the respective crops increased the grain yields of rice but there was decline in wheat yield. Partial substitution of chemical fertilizers with sesbanea green manure further improved the yield of rice and reversed the declining trend in wheat. The partial factor productivity, sustainable yield index, B: C ratio, soil organic carbon improved with the use of green manure with 50 or 75% recommended NPK fertilizers (Yadav *et al.,* 2000).

Crop Mixtures: When leguminous crops are used as a component in the mixture, they benefit other crops. During their growth, there is shredding of epidermal root cells, detachment of old root nodules and excretion of nitrogenous compounds from the plant roots into the soil. The nitrogen added to the soil by these means is utilized by other crops in the mixture. They are enabled to make better and vigorous growth. Cereals make better growth when grown with legumes than alone. They also make better growth than the other crops when grown mixed with legumes. The spreading fibrous roots of the cereals are possibly able to forage effectively and use the nitrogen excreted by the legumes better than the roots of other groups of plants.

Fertilizer formulation for crop mixtures can be derived using nutrient supplementation index (NSI). It is defined as the percentage usual uptake for that element by a sole crop of A that should be added to the mixture to meet the combined needs of intercropped A and B.

NSI = 100 [(Total uptake in mixture/Sole crop uptake) -1]

$$NSI_A = 100 \left(\frac{N_a + N_b}{N_A} \right) - 1 \quad \text{or generalized form of equation is}$$

$$NSI_A = \left(100 \frac{(N_a + N_b + \cdots + N_K)}{N_A} \right) - 1$$

N_A = Nutrient uptake by a sole crop of A per specified unit area of land

N_a = Nutrient uptake of A in the mixture of the same land area as sole cropped A

N_b = Nutrient uptake of B in the mixture of the same area of land as sole cropped A.

The ratio would be unity if the uptake in sole crops and mixtures were identical and in this case nutrient supplementation would not be needed. Thus a factor of -1 has been introduced into the equation for NSI. The conditions under which NSI would be valid are

1. The unit land area used for sole crop calculations must be the same as for the mixture.
2. The nutrient uptake rates must be determined at the same stage of development for crops in pure and mixed stands.
3. It does not matter whether the component species are at various stages of development, but all mixture samples must be taken from all components at the same time.
4. Calculations must be based on uptake per unit land area, not on uptake per plant.

The NSI information can be used for fertilizer formulation for crop mixtures under a given environment. It is essential to note that the NSI values refer to the amounts of nutrients taken up. The relation between them and amounts of fertilizer applied can be worked out for any given situation (Wahua, 1983).

QUESTIONS

Answer the following

1. Define integrated plant nutrient system according to food and agriculture organization.
2. What are the objectives of integrated nutrient management?
3. What are the steps of model integrated nutrient management?
4. What are the factors of fertilizer management in cropping system?
5. Explain nutrient management in crop mixtures.

Chapter 13

Soil and Fertilizer Management and Economics of Fertilizer Use

SOIL MANAGEMENT PRACTICES

- Good soil management is the basic for deriving greater benefits from the fertilizers used in crop production. Some of the good soil management practices are
- Good tilth is the first feature of good soil management. It means a suitable physical condition of the soil and implies in addition a satisfactory regulation of soil moisture and air.
- The maintenance of soil organic matter which encourages granulation is an important consideration of good tilth.
- Tillage operations and timings should be adjusted as to cause the minimum destruction of soil aggregates. Good tilth minimizes erosion hazards.
- The choice and sequence of adaptable crops or crop rotation are other very important considerations. These are related to climate, particularly rainfall and its pattern of distribution and the characteristics of the soil profile, including drainage and extent and duration of available soil moisture. A proper sequence of crop varieties greatly influences soil conditions. It is more realistic to evolve cropping patterns and land management practices according to land capability. Cropping patterns chosen and management practices adopted should aim at soil and moisture conservation for efficient nutrient and moisture utilization.
- In irrigated areas, special management practices become necessary to avoid salinity, alkalinity, water logging, leaching and the loss of plant nutrients. In rainfed areas special management practices include improving soil conditions to receive, retain and release more soil moisture., harvesting water to use as lifesaving irrigation or extending the cropping season when there is insufficient rainfall for raising crops, protecting the soil from degradation both in cropped and bare fields. Land shaping and leveling, mulching and the use of wind brakes and vegetative cover are the other major aspects.
- The productive capacity of the soil should never be allowed to diminish, but rather should be improved and maintained by providing adequate organic

manures and plant nutrients through fertilizers and by including legumes in the rotation and use of biofertilizers. Similarly provision should be made for irrigation facilities in semiarid and arid areas, adoption of different remedial measures against excessive salinity and alkalinity or acidity in humid areas, use of specific soil amendments to correct imbalances of plant nutrients and application of micronutrients where they are deficient.

- Economic plant protection measures against pests, pathogens and parasites including weeds should form part of the management practices in the cropping system. This can be achieved by following recommended cultural practices or by application of pesticides/fungicide/herbicides.

- The management practices adopted should be economically profitable and emphasis should be laid on maximizing sustained income rather than yields for the time being.

- An integrated land plan including all the above points and economically profitable should be developed for individual situations.

FERTILIZER MANAGEMENT

Factors influencing the time of fertilizer application

1. **Type of manures and fertilizers:** Bulky organic manures are applied much ahead (generally three weeks) of sowing so that they mix well, decompose and mineralize in the soil. Concentrated organic manures may be applied during sowing or planting. Nitrogen is highly mobile whereas phosphorus and potassium are relatively less mobile. Highly mobile nutrients should be applied in splits in smaller doses. Urea fertilizer is commonly used for top dressing. Green manures are added a week before seeding or planting.

2. **Soil type and soil water balance:** In sandy soils leaching loss of nutrients is greater and cation exchange capacity of the soil is lower. Readily soluble nutrients are partly applied immediately before sowing or at sowing. Later the remaining quantity is applied in splits as top dressing. In wet lands more splits are desirable. In dry lands entire quantity of nutrients required are added at sowing into the moist zone of the soil.

3. **Nature of the crop and cropping system:** Most of the crops require nitrogen throughout their growth period. The rate of uptake varies considerably. At the early stage of crop growth the rate of uptake is slow and gradually increases to a maximum and then declines to a minimum or nil. Uptake pattern also varies with the crop and management.

(a) Type of crop
Nitrogen is applied in two or more splits as per the duration and requirement of the crop, soil type and fertilizers selected.

Very short duration crops: Entire dose of nitrogen is applied in single application as basal dose

Medium duration crop: Two or three splits

Long duration crop: Three or four splits

Legumes: Starter dose of nitrogen, full dose of phosphorus and potassium is applied at sowing.

(b) Cropping situation
Dry land: basal dose alone

Irrigated: in splits

(c) Type of the economic product
In crops where the economic product is vegetative part nitrogen fertilizer application should be stopped much before the maturity *e.g.,* sugarcane, tobacco, sugar beet and potato.

In seed crops nitrogen application during seed development improves quality, number of seeds and weight of seeds. In cotton application of nitrogen at flowering improves the quality and yield. In fodder crops more number of splits improves protein content and succulence.

Crops require greater amount of phosphorus at early stage of growth. The entire dose may be applied as basal before or at sowing as phosphorus moves very little from the site of application.

Plants absorb potassium up to the harvesting stage but potassium fertilizers become available slowly. Potassium is less prone to losses. Therefore the entire quantity of potassium fertilizer may be applied at sowing. In warmer areas, light soils, sandy soils, under high nitrogen application, coastal regions and in poorly drained soils potassium application may be done in two to three splits.

Sulphur, calcium and magnesium fertilizers are applied before sowing or transplanting.

Elemental sulphur should be applied 2–3 weeks before sowing or planting into the moist soil.

Lime or dolomite lime stone can be applied much before sowing.

Micronutrients may be applied at sowing as soil application or as seed treatment.

Foliar application of micronutrients is done to alleviate the deficiency at early or at peak vegetative stage.

Boron is applied at flowering stage in sunflower.

Factors influencing the method of application

The method of application selected should reduce the loss of nutrient and easy accessibility to the plants.

Bulky organic and concentrated organic manures are generally broadcasted and incorporated during the final stage of tillage operation. Placement in seed rows or spot application is followed in wide spaced crops. *e.g.* chilli, brinjal, tobacco, castor, and redgram.

Nature of fertilizer

(a) **Powder form:** Single superphosphate is placed below the seed row or side dressed or incorporated. Basic slag, bone meal, rock phosphate is incorporated.

(b) **Granular fertilizers:** Urea briquette and prilled urea are broadcasted or placed.

(c) **Large granular fertilizers:** Large granular urea and urea super granules are placed in the reduced zone of wet soils.

(d) Liquid fertilizers are applied through irrigation water or injected into the soil. *e.g.*, liquid ammonia.

(e) **Gaseous form:** Ammonia – injected into the wet soil.

Mobility of nutrients: less mobile nutrients are placed near the roots.

Soil type

- **Acidic soils:** soluble phosphorus fertilizers are placed in bands while insoluble phosphorus fertilizers in water (soluble or insoluble in citric acid) are incorporated thoroughly into the soil.

- In soils rich in 2:1 type of clay minerals potassium is placed in bands.

- **Soil moisture status:** fertilizers should be placed into the moist zone of the soil in drylands using seed cum fertilizer drill or by appropriate techniques. Similarly for crops grown under residual soil moisture, entire recommended dose of fertilizer is placed in the moist zone in band below the seed row. Ammonical fertilizer nitrogen should be placed in the reduced zone of the paddy soil.

Nature of the crop and agrotechniques followed

In crops with fibrous root system and narrow spaced crops the roots are spread in the surface layer. Fertilizers are broadcasted and incorporated e.g. wheat, rice, finger millet, setaria.

- In crops with tap roots and widely spaced the fertilizers are placed in rings *e.g.*, cotton, chillies, tobacco *etc.*

- Fertilizers are band placed in a row at 5 cm from seed row in maize, jowar, bajra, sunflower.

- Based on the quantity: If the dose recommended is very small seed dressing can be followed. *e.g.*, molybdenum or cobalt.

Foliar Application of nutrients

Foliar fertilization is the practice of spraying major and micronutrient sources to the active growing foliage. This is absorbed through leaves. It is known as 'nonroot' or 'liquid manuring'. Earlier developed to alleviate iron and manganese chlorosis.

Suitability

- Sandy soils
- Saline, alkaline, acidic and water logged soils
- Mobile nutrients in plants

Methods

- Spray
- Dusting
- Overhead sprinklers

Mode of absoption

Stomata do not permit entry of water. Hence absorption is shifted to intercellular spaces of high frequency or ectodesmata. A thin cuticle and large surface area of the foliage favour penetration through the surrounding epidermal cell by the same channels described above. Addition of wetting agents or detergents to the spray solution increases the efficiency of nutrient absorption by the foliage. In addition to these the nutrients are also absorbed by the foliage through diffusion and exchange.

Urea: Nitrogen absorbed by passive process. Nutrient cations are generally absorbed while least absorbed is anions.

Rate of nutrient absorption into plant tissues.

Table 39. Time taken for 50 percent absorption

Nutrient	Time
Nitrogen: urea	½ to 2 hrs
Phosphorus	5–10 days
Potassium	10–24 hrs
Calcium	1–2 days
Magnesium	2–5 hrs
Sulphur	8 days
Zinc	1–2 days
Manganese	1–2 days
Iron	10–20days
molybdenum	10–20 days

Urea: Time taken for absorption is 1–4 hrs in citrus, apple and pineapple, 24 hrs in sugarcane.

Time of application: perennial fruit trees any time during the growing season. Stage of foliar application of nutrients should coincide with panicle initiation in cereals, 35–45 days after sowing in fibre and oil seed crops and 30–40 days after sowing in vegetable crops.

Time of the day for spraying: Late evening (after 6.00 pm) and early morning (before 9.00 am).

Temperature of the atmosphere: 18 °C–29 °C, 21 °C is ideal.

Humidity should be more than 70 per cent.

Wind speed should be less than 8 km ph.

Table 40. Fertilizers suitable for foliar fertilization

Nutrient	Fertilizer	Concentration (%)
Nitrogen	Biuret free urea (<0.2%)	21–44
	Ammonium polyphosphate	10–21
	Ammonium thiosulphate	12
	Potassium nitrate	13.75
Phosphorus	Ammonium polyphosphate	33–52
	Orthophosphates (liquid)	4–18
Potassium	Potassium nitrate	44.5
	Potassium thiosulphate	25
	Potassium hydroxide	6–18
	Potassium sulphate (17.6% S)	54
Calcium	Calcium sulphate	23
	Calcium nitrate	21
Magnesium	Magnesium sulphate	10
	Magnesium nitrate	6.3
Sulphur	Ammonium thiosulphate	26
	Ammonium sulfate solution	9
	Potassium thiosulphate	17
	Potassium sulphate	17–18
	Magnesium sulphate	12–13
	Calcium sulphate	15–18
	Zinc sulphate	18
	Manganese sulphate	13–18
	Iron sulphate	19
	Copper sulphate	13
Zinc	Zinc sulphate	36
	Metalosates	6.8
Manganese	Manganese sulphate	25–28
	Metalosates	5.0

Nutrient	Fertilizer	Concentration (%)
Iron	Ferrous sulphate	20
	Ferrous ammonium sulphate (17% S and 10% N)	14
Copper	Copper sulphate (12% S)	35
Boron	Solubar	20.5
Molybdenum	Ammonium molybdate	54
	Sodium molybdate	38

Chelates are not suitable because of their large molecular structure. Concentration of the spray solution depends on

- **Age of the crop:** Young crop – diluted, Aged crop- higher concentration
- Type of the sprayer used:
  Low volume: Higher concentration
  High volume: Lower concentration

Table 41. Concentration of fertilizer nitrogen

Crop	High volume (%)	Low volume (%)
Wheat	3–6	10–30
Paddy	2–4	10–20
Maize	2–4	10–20
Bajra/jowar	3–6	10–20
Sugarcane	4–6	15–30
Cotton	1–2	10–15
Oilseeds	1.5–2	10–15

Table 42. Quantity of fertilizer nitrogen (kg) that can be supplied per hectare

Type of sprayer	Optimum quantity	Maximum quantity
High volume	10	20
Low volume	20	40

Advantages
- Nutrients are directly absorbed- efficiency is more
- Crop response is quicker
- Overcomes transformation of nutrients when applied to soil – by chemical transformations, fixations, immobilization, leaching *etc.* which reduce the efficiency of applied fertilizers.

- More effective during low soil temperature- Root activity and ion activity is reduced at low soil temperature which reduces root absorption.
- More effective under soil moisture deficient situation.
- Under situations of root injury or damage by insect pests or diseases.
- Pesticides and fungicides can be sprayed along with foliar spray depending upon the compatibility
- Hidden hunger of nutrients may be alleviated.

Limitations
- Entire requirement of nutrients cannot be provided through foliar spray.
- Biurette (NH_2 CO NH CO NH_2) in urea should be less than 0.2 per cent.

Fertigation
Fertigation is the application of fertilizers through irrigation water.

Criterion for selection of fertilizers
1. Soluble and emulsifiable in water.
2. Should not adversely react with salt and other chemicals present in irrigation water.
3. Should not change the pH of irrigation water.
4. Should not leach out from the root zone.
5. Should not be corrosive or clogging in the irrigation system.

Fertilizers

Nitrogen sources: Nitrate nitrogen sources are more suitable. They may be mixed with ammonium sulphate, urea, ammonium nitrate, calcium nitrate when bicarbonate levels are low in irrigation water. Ammonia or ammonium forms clog the system.

Phosphorus sources: Inorganic phosphorus sources cause clogging problem if the irrigation water contains higher calcium and magnesium. Phosphoric acid, glycerophosphate (organic phosphate) and orthophosphates are tried. Glycerophosphate and orthophospharic acid are found suitable in most cases.

Potassium sources: Common potassium sources are potassium chloride, potassium sulphate and potassium nitrate. They are water soluble and not readily leached out from the root zone.

Micronutrients: Chelates of iron, copper, zinc and manganese can be applied in chelated forms or sulphate salts. Since quantity required is small precise metering is required.

Maintenance

Clogging caused by calcium or magnesium carbonates can be overcome by periodic injection with HCl or H_2SO_4.

Clogging caused by microbial growth (bacterial slime or fungi) can be overcome by injecting chlorine or sodium hypochlorite at 1 ppm continuously or slug treatment at 10–20 ppm at intervals as required.

Table 43. Fertilizers suitable for fertigation and nutrient content

Sources	Solids	Saturated liquid
	$N: P_2O_5: K_2O$ (%)	$N: P_2O_5: K_2O$ (%)
Urea	46:0:0	21:0:0
Ammonium nitrate	33:0:0	21:0:0
Ammonium sulphate	21:0:0	10:0:0
Phosphoric acid	0:0:0	61:0:0
Monoammonium phosphate	12: 61:0	4:18:0
Diammonium phosphate	18:46:0	7:25:0
Potassium chloride	0:0:62	0:0:15
Potassium nitrate	13:0:46	4:0:12
Potassium sulphate	0:0:50	0:0:6
Monopotassium phosphate	0:52:34	0:10:7

Economics of fertilizer use

Use of fertilizer for increased crop production depends entirely on its economics. This is usually done by reporting response per unit area or per unit nutrient applied. Cost benefit ratio is the commonly used economic indicator of fertilizer use. The calculations are based on the extra produce at the market price and deducting the cost of fertilizer only at prevailing market price.

$$P = R - E$$

Where P is additional profit, R is the value of additional return, E is additional expenses incurred consequent fertilizer use.

Additional income (P) due to fertilizer use consists of number of components, some of which are as follows:

(a) Value of extra grain or main produce

(b) Value of secondary or byproduct *e.g.*, Extra stover/stalk/straw yield due to fertilizer use

(c) Residual effect, if significant, leading to increased yield of the subsequent crop and

(d) Premium or discount in the price of the produce due to use of fertilizer

Regarding expenses (E) it is to be noted that the application of fertilizer to crop usually entails additional expenses by way of extra irrigation, extra interculture, additional plant protection measures, *etc.* It would also be correct to say that the additional yield consequent to the use of fertilizer is the combined result of the effect of fertilizer and these additional inputs. Any attempt to properly evaluate the economics of fertilizers only should deduct the increase due to other associated inputs.

Cost of production of a crop has the following two well recognized components

- **Fixed costs:** The cost of which does not vary with the level of yield or output like cost of seed, preparatory tillage operations, post sowing intercultivation *etc.*
- **Variable cost:** The cost which varies directly with the level of produce, like cost of harvesting, threshing, winnowing, marketing *etc.*

When cost- benefit ratio is worked out without considering value of secondary produce and variable cost of cultivation, it is known as Gross Cost Benefit Ratio and when value of secondary produce and variable cost are considered the ratio is known as Net Cost Benefit Ratio.

Cost Benefit Ratio (CBR) for fertilizer application (economics of fertilizer use) is a variable factor, changing directly with value of extra main and secondary produce and fertilizer cost.

Another ratio commonly used in working out economics of fertilizer use is Incremental Cost Benefit Ratio (ICBR) (or added cost added return). It is a decision making tool which can be applied in evaluating profitability of using certain doses of fertilizers for enhancement of crop yields. In ICBR analysis, the incremental costs of applying fertilizers and the incremental benefits through additional farm output are compared. As per FAO standard the value of ICBR should be more than 1: 2.5. To work out ICBR the control plot with no fertilizer input is required.

Opportunity cost principle or decision rule has been applied to decide which crops to fertilize at what level. Profits are at maximum when each unit of capital, labour, land and management is used in the place where it adds most to returns.

Fertilizer use on different soils of the same farm: fertilizer must be applied to soils which give most profit per unit expenditure on fertilizer.

Optimum rates of several nutrients: in making decisions on fertilizer in most instances farm manager is not concerned with a single nutrient but with optimal rates of several nutrients. Fertilizer nutrients quite often interact to produce an added crop responses of the nutrients used separately. A farm manager is interested in utilizing these interactions and realizing their fullest profit advantage.

Substitution of fertilizer to land: what is the least cost combination of land and fertilizer to use in producing a given amount of product?

Residual or carry over effect: fertilizer cost is incurred in one period and part of the returns is received in a later period, future returns should be discounted back to the time when the cost is incurred if a valid comparison of costs and returns is to be made.

Factors other than yields affecting the most profitable rate: fertilizers may also add to quality of the product. If wheat, for example should be bought on the basis of protein content, farmers could afford to increase the yields of high protein grain through increased fertilization.

Barley for feed and malting: If barley is grown for feed, the most profitable fertilization rate can be determined without difficulty. Using added cost and added returns principle. If it is for malting, the malting premium is considered.

QUESTIONS

Answer the following

1. What are the good soil management practices?
2. Explain the factors influencing the time of fertilizer application.
3. Explain the factors influencing the method of fertilizer application

Short answers

1. How the foliar applied nutrients are absorbed?
2. What is the optimum crop stage for foliar application of nutrients?
3. What is the appropriate time of the day for foliar spray of nutrients?
4. What are the advantages of foliar application of nutrients? What are the criterion for selection of fertilizers for foliar nutrition?
5. What are the components of additional income and additional expenses of fertilizer use in crop production?
6. What are the components of cost of production?

Define

1. Incremental cost benefit ratio
2. Cost benefit ratio
3. Opportunity cost principle of fertigation
4. Gross cost benefit ratio
5. Net benefit cost ratio

CHOOSE THE CORRECT ANSWER

1. The time taken for 50 per cent absorption of foliar applied nutrients is lowest in
   (a) Urea nitrogen
   (b) Phosphorus
   (c) Potassium
   (d) Zinc

2. The quantity of nitrogen that can be supplied per ha with low volume sprayer is
   (a) 20–40kg per ha
   (b) 10–20 kg per ha
   (c) 40–50 kg per ha
   (d) 50–60 kg per ha

3. Most suitable phosphorus fertilizer for fertigation
   (a) Single superphosphate
   (b) Orthophosphoric acid
   (c) Bone meal
   (d) Diammonium phosphate

4. Clogging by carbonates of calcium and/or magnesium can be overcome by
   (a) Sulphuric acid
   (b) Potassium di hydrogen phosphate
   (c) Calcium hydroxide
   (d) Sodium hypochlorate

5. Clogging in drip system can be overcome by injection
   (a) Sodium hypochlorate
   (b) Potassium hypochlorate
   (c) Magnesium hypochlorate
   (d) Calcium hydroxide

6. Urea fertilizer for foliar application should have biuret content
   (a) <0.2%
   (b) 2%
   (c) 3%
   (d) 4%

Chapter 14

Nutrient Management in Problem Soil

Salt affected soils occupy 962.2 m ha globally which accounts for 33% of the arable land. In India 8.373 m ha is affected with salts. Of which 2.359 m ha are alkali soils, 3.829 m ha are saline soils and 2.185 m ha coastal saline soils. Reclamation of these soils provides additional area for food grains production and employment.

ALKALI SOILS

Alkali soils have electrical conductivity <4.0 dSm^{-1}, pH > 8.2 and ESP > 15.

Nutrient related problems

- **Deficiency of calcium:** Since nearly all the soluble and exchangeable calcium is precipitated as insoluble $CaCO_3$.
- Excess of sodium which is toxic per se to the plants and causes imbalance due to antagonistic effect of potassium and calcium nutrition.
- Toxic concentrations of HCO_3^- and CO_3^- ions.
- Decreased solubility and availability of micronutrients *viz.* zinc, iron, manganese, due to high pH, $CaCO_3$, soluble CO_3^- and HCO_3^-
- Increased solubility and accumulation of certain elements like fluorine, selenium and molybdenum at toxic levels in plants that may affect crop yield and/or health of animals feeding on such crops.. Increased concentration of fluorine can cause bone diseases in animals. Selenium toxicity cause Degnala disease, which affects skin and hair, is quite common among cattle fed on fodder and straw of crops raised on alkali soils.

Low activity of useful microbes due to high pH and excess soluble sodium.

Management practices to lower the exchangeable sodium per cent (ESP)

Amendments are used to lower the ESP may directly provide calcium required to displace sodium from the exchange complex. There are three types of amendments.

Chemical amendments: Gypsum and calcium chloride, pyrites and mineral sulphur.

Organic amendments: Rice straw, rice husk, poultry droppings, groundnut and sunflower hulls; farmyard manure, compost or green manure, tree leaves, saw dust *etc.* They release carbon dioxide upon decomposition which dissolves calcite and release calcium.

Organic and inorganic amendments should be used together *e.g.* 20 t FYM and gypsum.

Material having wide C: N ratio is less effective as compared to that of narrow C: N ratio. They may cause nitrogen deficiency.

Industrial by-products: Phosphogypsum, pressmud, molasses, acid wash and effluents from milk plants may be used to provide soluble calcium directly or indirectly by dissolving native calcium sources. Care should be taken not to use materials containing high fluorine *viz.* phosphogypsum.

Activated pressmud (pressmud treated with H_2SO_4) is rich in organic matter and other nutrients.

Gypsum

Gypsum containing 70% $CaSO_4$ $2H_2O$ at 10-15 t per ha is required which is approximately 50% of the gypsum requirement by Schoonovers method to replace sodium from the soils to a depth of 15cm. Gypsum is applied 2-3 days before sowing or planting of crops. Pyrite requirement is 85% of gypsum requirement. Nine to thirteen tones of pyrites are required per ha. Gypsum used should have fineness of 2mm while pyrites of 5 mm. Continuous application during first three years is sufficient. If the irrigation water contains more sodium repeated application is required. The quantity of the material required depends on the clay content of the soils. Larger the clay or heavy soils require more gypsum than sandy soils.

Nutrient management

Nitrogen: Available nitrogen is low in alakali soils as these soils are low in organic matter content.

Losses due to volatilization- frequent cycles of wetting and drying leads to loss of nitrogen through denitrification.

Ammonium sulphate, ammonium nitrate, calcium ammonium nitrate, monoammonium phosphate and diammonium phosphate, muriate of potash, sulphate of potash are suitable for alkali soils.

Alkali soils need more nitrogen. Twenty five per cent more nitrogen is required over the recommended nitrogen levels. Losses can be mitigated by split application. Foliar application of nitrogen is more efficient method as it can save 40–60 kg N per ha. Green manuring with sesbania enhances nitrogen supply to crops.

Phosphorus: Alkali soils contain large amount of extractable phosphorus. sodium carbonate or bicarbonate react with insoluble calcium phosphate forming soluble sodium phosphate with steep increase in Olsen's extractable phosphorus. Phosphorus management should be on soil test basis.

Potassium: High amount of sodium and low calcium results in sodium accumulation in plants. As a consequence potassium uptake by the plants is adversely affected. This is known as sodium induced potassium deficiency. Under situations of medium to high levels of exchangeable potassium it is suggested to correct the Ca: Na balance by using amendments rather than potassium application.

Zinc: Alkali soils contain medium to high amounts of total zinc (40–100 mg kg^{-1} of soils) which is comparable with nonalkali soils. High pH, presence of $CaCO_3$, high soluble phosphorus, toxic concentrations of CO_3 and HCO_3 and low organic matter render Zn deficiency in alkali soil. They contain less than 0.6 mg kg^{-1} DTPA extractable zinc. Application of 10–15 t of gypsum +10-20 kg $ZnSO_4$ ha^{-1} is enough to meet the zinc requirement of rice- wheat cropping. Continuous application of zinc using zincated urea, zinc enriched farmyard manure, or dipping rice seedlings in 3% ZnO solution alleviate zinc deficiency. Foliar application using 0.5% zinc sulphate solution at an interval of 10 days is superior over soil application.

Iron: Due to excess of CO_3^- and HCO_3^- iron is transformed into unavailable form. Iron chlorosis is observed in rice nurseries, sugarcane and tomato, soybean.

Iron deficiency can be alleviated by green manuring, Keeping the soil under flooded condition, and pyrite application. Iron is immobile element in plants. Hence foliar application can only supplement the other management practices. Foliar spray of 0.3% ferrous sulphate solution is used.

Manganese: High soil pH and $CaCO_3$ adversely affects the availability of manganese to plants. In wheat growing areas it is a limiting factor. Manganese leaches to lower layers under submerged condition.

Deep ploughing, repeated sprays with $MnSO_4$ are the management practices. Soil application of manganese will not be useful.

Table 44. Crops tolerant to alkalinity

Tolerant	Moderately tolerant	Sensitive
Karnal grass	Dhaincha	Cowpea
Paragrass	Wheat	Bengalgram
Rhodes grass	Berseem	Lentil
Bermuda grass	Barley	Groundnut
Rice	Mustard/Raya	Peas
Sugarbeet	Sugarcane	Maize
	Pearlmillet	

Saline soils

Saline soils contain high concentration of soluble salts *viz.* chlorides and sulphates of sodium, calcium, magnesium. EC > 4, dS/m, pH 8.2, ESP <15. Saline soils are in flocculated state and hence fairly permeable. Plant growth in such soils is affected by high osmotic effects of salts, toxicity of soluble ions like sodium, chlorine and boron and lesser availability of other essential nutrients.

Reclamation: Leaching excess salts is the only option. Availability of abundant fresh water and drainage are the requirements.

Nitrogen: Deficiency of nitrogen is wide spread in saline soils due to poor vegetation growth and lack of organic matter. Large fraction of applied nitrogen is lost in gaseous form under high salinity. High concentration of salts inhibits nitrification and resultant $NH_4^+ - N$ accumulation. Hence the plants which absorb nitrate nitrogen will show deficiency of nitrogen even if soils have adequate nitrogen.

Availability of phosphorus and potassium: Availability of phosphorus increases up to a moderate level and there after decreases. Saline soils are medium to high in available potassium. But plants grown under high salinity may show potassium deficiency due to antagonistic effect of sodium and calcium on potassium absorption or disturbed Na/K or Ca/K ratio. Under such conditions application of potassium may give increased yield.

Management of nutrients

Judicious application organic manures and fertilizers would be inevitable to ensure stable crop yield. Fertilizers with low salt index should be used. Band placement of phosphorus, deep placement of nitrogen fertilizers and foliar application of nitrogen are essential.

Nitrogenous fertilizers are required in larger quantity over the recommended dose. Wheat responded to 160 kg N per ha, Barley to 120 kg N per ha. Pearl millet to120 kg N per ha and cotton to 80 kg N per ha. Split application of nitrogen and deep placement of nitrogen should be followed. Sulphur coated urea was found superior to lac coated urea.

Phosphorus application helps in increasing the crop yields directly by providing phosphorus to the crops or by decreasing the absorption of toxic elements *viz.* fluorine or chloride. Yield of wheat and mustard increased with the application of 30 kg P_2O_5 ha^{-1}

Potassium application may increase the crop yields either by directly supplying potassium or by improving its tolerance to sodium, calcium and magnesium. However it is difficult to exclude sodium from the plant by use of potassium fertilizers.

Micronutrient deficiencies in saline soils are location specific. Hence soil test based applications may be followed. In rice- barley cropping system application of 10 kg zinc sulphate was found beneficial.

Table 45. Crops tolerance to salinity

Tolerant	Moderately tolerant	Sensitive
Barley	Wheat	Pulses
Sugarbeet	Sugarcane	Linseed
Safflower	Maize	
Cotton	Sunflower	
Mustard	Castor	
Paragrass	Oats	
Chillies	Tomato	
Finger millet		

Acid soils

Acid soils are characterized by low pH (generally < 5.5). Crop growth is mainly affected by excessive amounts of aluminium, manganese and copper which may attain toxic levels.

Liming is recommended according to lime requirement which varies with soil pH, clay content, CEC and climate. The approximate quantity of finely ground lime stone required to raise the pH to 6.5 in the plough sole layer (15 cm depth) is given in the Table 46.

The liming material should be evenly broadcasted and mixed into the soil a few weeks in advance to sowing the crops to complete the exchange process. A proper moisture content in soil is essential. Liming is advocated once in five years. However, soil pH needs to be checked out once in a year.

The present emphasis is on the determination of aluminum saturation and recommendation of lime based on the neutralization of exchangeable Al. Nevertheless, exchangeable H too demands attention in this regard. Since most of the acidic red soils (predominant group of red soils) are in the pH range of 5.2–6.0 the range in which both Al^{+++} and H^+ are active proton donors.

Table 46. Lime requirement of acid soils.

Soils	Lime stone (t/ha) required to raise the pH to 6.5		
	Initial soil pH 4.0	Initial soil pH 4.5	Initial soil pH 5.5
Warm humid plains			
Sand and loamy sands	3.75	2.5	1.25
Sandy loams		5.0	2.5
Loams and silty loams		8.75	5.0
Clay loams		12.5	7.0
Cool temperate hilly region			
Sands and loamy sands	7.5	5.5	2.5
Sandy loams		7.5	5.0
Loamy and silt loams		11.25	7.5
Clay loams		15.0	8.75
Valley soils			
Organic water logged soils	22.5	17.5	11.5

Table 47. Crop response to fertilizer in presence of lime in acid soils

Highly responsive	Medium responsive	Non responsive
Redgram	Bengalgram	Paddy
Soybean	Lentil	Small millets
Cotton	Peas	mustard
	Maize	
	Sorghum	

Liming is an important agronomic practice to sustain soil health. Liming inactivates iron, aluminum and manganese and reduces phosphorus fixation. Stimulates the microbial activity, enhances the fixation of atmospheric nitrogen and helps mineralization of organic matter.

Besides lime, industrial wastes *viz.* steel mill slag, blast furnace slag, lime sludge from paper mills, cement klin wastes, precipitated $CaCO_3$ can be used. Application of paper factory sludge (pH 5.52, 19.3% organic carbon, 0.81% N, 0.12% P and 0.44% K) along with fly ash (pH 8.47, 0.03% organic carbon, 0.05% N, 0.03% P and 0.18% K) and recommended fertilizers produced highest yield of rice as well as higher residual effects on the subsequent crops.

Aluminium tolerant crops/varieties such as paddy, finger millet and other minor millets may be selected in acid soils. Tolerant plant species can withstand the high concentrations of aluminium or manganese by effective translocation and compartmentalization, while species with avoidance mechanisms will exclude aluminium and manganese from sensitive sites, exclusion from uptake by the root induced changes in the rhizosphere and/or efficient in nutrient acquisition through rhizosphere microorganisms association or mycorrhizae.

Nutrient management

Liming is basic for nutrient management in acid soils. Liming increases the availability of nitrogen. Acid forming fertilizers should be avoided. Calcium ammonium nitrate was found better in wheat grown in acid soils.

Nitrogen loss in acid soils can be minimized by split application. Leaching losses of N applied in a single dose amounts to 50 to 60% while under three split applications it was 11–33%. In alkali soils volatization loss of nitrogen is more (50%) while in acid soils volatilization losses are very less (6%).

Use of crop residues improves the efficiency of urea by 52.5%. Cured urea application is beneficial. Use of materials to reduce dissolution of urea *viz* neem coating, rock phosphate coating are beneficial. Modified urea materials *viz.* urea super granules, large granular urea may be used to reduce nitrogen losses.

Selection of suitable phosphatic fertilizer plays key role in phosphorus management in acid soils. Avoid water soluble phosphorus sources *viz.* single super phosphate, triple super phosphate or diammonium phosphate. Citrate soluble or citrate insoluble sources are economical. mixtures of Massurie rock phosphate or Udaipur rock phosphate and single super phosphate are appropriate in several crops and cropping systems.

Application of lime takes care of calcium requirement of crops however excess of liming set backs the potassium uptake by the crops.

Micronutrients: Response of crops to iron, manganese, zinc and copper is very much limited in the acid soils. However rice responds to zinc as its behavior is indifferent. Climatic factors, high levels of phosphorus, growing sensitive genotypes, continuous use of urea fertilizer favour the response of crops to zinc application.

Acid soils are low in molybdenum and boron. Hence the crops respond to application of these nutrients.

Acid sulphate soils

Acid sulphate soils account for 12.5 m ha in the world and 3 m ha in India. They are localized in the low lying coastal lands which are generally rich in pyrite and on oxidation form sulphuric acid. Acid sulphate soils have pH < 4.0. The crops in these soils suffer mainly because of aluminium toxicity and nutrient deficiency especially phosphorus.

Water management is the key to soil management in acid sulphate soils.

Root development is often restricted in acid sulphate soils unless water is available for irrigation or to maintain a high water table artificially. As flooding alleviates acidity, rice is the only choice to be grown in acid sulphate soils.

Total reclamation of acid sulphate soils can be slow and expensive. Reclamation is relatively easier or less costly on the soils which do not have reserves of pyrite or jarosite close to the surface or high values of exchangeable aluminium. On such soils moderate application of lime will significantly reduce soluble aluminium levels and enable crops to respond to conventional fertilizer application without raising the pH above 4.5.

QUESTIONS

CHOOSE THE CORRECT ANSWER

1. Degnala disorder in cattle is due to feeding fodder crop grown in alkali soils rich in
   (a) Fluorine   (b) Sodium
   (c) Selenium   (d) Calcium

2. The cattle fed with fodder crop grown in alkali soils suffer from bone diseases. The cause is accumulation of excess_____ in the fodder.
   (a) Fluorine   (b) Chlorine
   (c) Selenium   (d) Molybdenum

3. The total zinc is more, but available zinc is less in
   (a) Saline   (b) Alkaline
   (c) Acid   (d) Acid sulphate

4. Excess of sodium in alkali soils adversely affect the uptake of
   (a) Phosphorus   (b) Zinc
   (c) Potassium   (d) Iron

5. The crop suitable for alkali soils
   (a) Rice   (b) Mmaize
   (c) Groundnut   (d) Peas

6. Gypsum application is recommended for reclamation of
   (a) Saline soils   (b) Acid soils
   (c) Acid sulphate soils   (d) Alkali soils

7. Potassium uptake is hindered by sodium in
   (a) Alkali soils   (b) Saline soils
   (c) Acid soils   (d) Submerged soils

8. Volatilization losses of nitrogen is more in
   (a) Acid soils
   (b) Alkali soils
   (c) Acid sulphate soils
   (d) Sandy soils

9. The crop suitable for acid sulphate soils is
   (a) Rice
   (b) Safflower
   (c) Sugarcane
   (d) Sweet potato

10. The major problem in acid sulphate soils is the toxicity of
    (a) Iron
    (b) Aluminium
    (c) Manganese
    (d) Chlorine

11. Nitrification is inhibited in
    (a) Acid soils
    (b) Saline soils
    (c) Dry soils
    (d) Submerged soils

12. Pyrite is used as an amendment for reclamation of
    (a) Saline soils
    (b) Alkali soils
    (c) Acid soils
    (d) Acid sulphate soils

Answer the following

1. What are the nutrient related problems in alkali soils?
2. What are the management factors for reducing exchangeable sodium in alkali soils?
3. Explain the management of NPK in alkali soils?
4. Explain zinc, iron and manganese management in alkali soils.
5. What are the nutrient related problems in saline soils?
6. What are the specific nutrient management practices in saline soils?
7. Explain the nutrient management for crops grown in acid soils.
8. What are the advantages of liming in acid soils?

References

Barber S.A., Munson R.D. and Dancy W.B.1985. Production, marketing and use of potassium fertilizers. Pp377-410. In O.P.Englestad (Ed.), Fertilizer Technology and Use. Soil Science Society of America, Madison, Wisc.

Becket P. H. T., 1964, Studies on potassium II- The immediate Q/I relationship of labile potassium in soil. J. Soil Sci. 15:9

Chandler, W.H. *et. al.* 1932. Proc. Ame. Soc. Hort. Sci. 28, 556-560

Grossenbacher, (1916) and Floyd (1917) In Hand book of copper compounds and applications-Technology and Engineering, Wayne Richardson, 1997 (Ed.) https://books.google.co.in. books

Horstmann,1911. In Sulphur fertilization in Indian agriculture –A Guide Book, FDCO, New Delhi, Ed.by HLS Tandon, 101, 1995.

Liebig, 1855, In soil fertility evaluation and control by C. A. Black, 1992, Lewis Publishers, Boca Raton.

Macy, 1936, In soil fertility evaluation and control by C. A. Black, 1992, Lewis Publishers, Boca Raton.

Mitscherlich,1909, In soil fertility evaluation and control Ed. by C. A. Black, 1992, Lewis Publishers, Boca Raton.

Wahua, T.A.T. 1983, Nutrient uptake by intercropped maize and cowpeas and a concept of nutrient supplementation index (NSI). Experimental Agriculture, Volume 19, Issue 3, pp. 263-275. Cambridge.org. agris.fao.org (published on line by Cambridge University Press).

Yadav, R. L. 1998, Fifty years of Agronomic Research in India. Indian Society of Agronomy, IARI, New Delhi.

Appendices

Equivalent acidity is the number of kg of calcium carbonate required to neutralize the acidity produced by application of 100 kg of fertilizer.

Fertilzer	Equivalent acidity
Ammonium sulphate	110
Urea	80
Ammonium nitrate	60
Ammonium sulphate nitrate	85
Ammonium chloride	128
Anhydrous ammonia	148

Equivalent basicity is the basicity produced by 100 kg of fertilizer in terms of the effect produced by the number of kgs of calcium carbonate.

Fertilizer	Equivalent basicity
Calcium nitrate	21.5
Sodium nitrate	29
Calcium cyanamide	63
Single superphosphate	19.5
Dicalcium phosphate	25

Salt index (SI) is the ratio of the increase in osmotic pressure produced by the material in question to that produced by the same weight of sodium nitrate based on a relative value of 100.

Fertilizer	SI per unit of plant nutrient
Sodium nitrate	6.06
Potassium nitrate	5.34
Ammonium sulphate	3.25
Ammonium nitrate	2.99
ᴐoammonium phosphate	2.45
	1.62
ᴐum phosphate	1.61
ᴐhosphate	ᴐ 39
ᴐhate ᴐ	0.21
	1.97
	1.94
	1.58
	0.85

Specific leaf N is the leaf N per unit leaf area required for expansion. If N uptake and allocation to leaves is sufficient to maintain SLN above 1.0 g per m^2 then leaf expansion proceeds at its maximum potential rate.

Minerals containing micronutrient

S. No.	Micronutrient	Minerals
1.	Boron	Tourmaline
2.	Copper	Chalcopyrite, Chalcocite, Bornite
3.	Iron	Hematite, Siderite, Olivine, Pyrite, Goethite, Magnetite, Limonite
4.	Manganese	Pyrolusite, Hausmannite, Manganite, Rhodochrosite, Rhodonite
5.	Molybdenum	Molybdenite
6.	Zinc	Augite, Hornblande, Biotine, Spharalite, Smithsonite, Hemimorphite, Franklenite, Willemite